MASCULINITIES IN FORESTS

Masculinities in Forests: Representations of Diversity demonstrates the wide variability in ideas about, and practice of, masculinity in different forests, and how these relate to forest management.

While forestry is widely considered a masculine domain, a significant portion of the literature on gender and development focuses on the role of women, not men. This book addresses this gap and also highlights how there are significant, demonstrable differences in masculinities from forest to forest. The book develops a simple conceptual framework for considering masculinities, one which both acknowledges the stability or enduring quality of masculinities, but also the significant masculinity-related options available to individual men within any given culture. The author draws on her own life, building on her long-term experience working globally in the conservation and development worlds, also observing masculinities among such professionals. The core of the book examines masculinities, based on long-term ethnographic research in the rural Pacific Northwest of the US; Long Segar, East Kalimantan; and Sitiung, West Sumatra, both in Indonesia. The author concludes by pulling together the various strands of masculine identities and discussing the implications of these various versions of masculinity for forest management.

This book will be essential reading for students and scholars of forestry, gender studies and conservation and development, as well as practitioners and NGOs working in these fields.

Carol J. Pierce Colfer is a Senior Associate at the Center for International Forestry Research (CIFOR) and Visiting Scholar at Cornell University's Southeast Asia Program, Ithaca, New York, USA. She is author/editor of numerous books, including co-editor of *The Earthscan Reader on Gender and Forests* (Routledge, 2017) and *Gender and Forests: Climate Change, Tenure, Value Change and Emerging Issues* (Routledge, 2016).

The Earthscan Forest Library

This series brings together a wide collection of volumes addressing diverse aspects of forests and forestry and draws on a range of disciplinary perspectives. It is aimed at undergraduate and postgraduate students, researchers, professionals, policy-makers and concerned members of civil society.

Forest Ecosystem Management and Timber Production
Divergence and Resource Use Resilience
Russell Warman

A New Era for Collaborative Forest Management
Policy and Practice Insights from the Collaborative Forest Landscape Restoration Program
Edited by William Butler and Courtney Schultz

Forest Conservation and Sustainability in Indonesia
A Political Economy Study of International Governance Failure
Maxton Bernice-Lee

Masculinities in Forests
Representations of Diversity
Carol J. Pierce Colfer

For more information about this series, please visit: www.routledge.com/books/series/ECTEFL

MASCULINITIES IN FORESTS

Representations of Diversity

Carol J. Pierce Colfer

Routledge
Taylor & Francis Group
LONDON AND NEW YORK

earthscan
from Routledge

First published 2021
by Routledge
2 Park Square, Milton Park, Abingdon, Oxon OX14 4RN

and by Routledge
52 Vanderbilt Avenue, New York, NY 10017

Routledge is an imprint of the Taylor & Francis Group, an informa business

British Library Cataloguing-in-Publication Data
A catalogue record for this book is available from the British Library

Library of Congress Cataloging-in-Publication Data
Names: Colfer, Carol J. Pierce, author.
Title: Masculinities in forests : representations of diversity / Carol J. Pierce Colfer.
Description: Abingdon, Oxon ; New York, NY : Routledge, 2021. | Includes bibliographical references and index.
Identifiers: LCCN 2020022564 (print) | LCCN 2020022565 (ebook) | ISBN 9780367403898 (hardback) | ISBN 9780367417000 (paperback) | ISBN 9780367815776 (ebook)
Subjects: LCSH: Forests and forestry—Social aspects. | Forest management—Social aspects. | Masculinity. | Sex role in the work environment.
Classification: LCC SD387.S55 C65 2021 (print) | LCC SD387.S55 (ebook) | DDC 363.37/9—dc23
LC record available at https://lccn.loc.gov/2020022564
LC ebook record available at https://lccn.loc.gov/2020022565

ISBN: 978-0-367-40389-8 (hbk)
ISBN: 978-0-367-41700-0 (pbk)
ISBN: 978-0-367-81577-6 (ebk)

Typeset in Bembo
by Apex CoVantage, LLC

This book is dedicated to those many men in the world's forests

- *who conduct their lives with a gentle, loving and respectful spirit;*
- *who use their strength to protect and improve the lives of those around them;*
- *whose intelligence contributes to a better and more equitable world for all.*

CONTENTS

FIGURES

TABLES

BOXES

CONTRIBUTORS

Carol J. Pierce Colfer is a cultural anthropologist much of whose work since 1973 has focused on peoples living in forests. She became interested in gender issues earlier still, while in high school (in the early 1960s), and has retained that interest ever since.

Sufiet Erlita is in charge of CIFOR's library and provided valuable help in analyzing the quantitative materials from a March 2019 revisit to East Kalimantan for Chapter 7.

Susan Paulson has spent decades exploring farm and forest management in Andean and Amazonian contexts. She also learns about masculinities, gender and environment by facilitating workshops and collaborative publications, as well as graduate teaching in South America, Europe and currently at the University of Florida.

Andrew Balan Pierce is an artist, entrepreneur and martial arts expert who splits his time between Bali, Indonesia and Portland, Oregon and has long experience with the Kenyah of East Kalimantan, being half Kenyah himself. His contribution to the book has added a masculine interpretation in Chapter 7.

ACKNOWLEDGEMENTS

Throughout the writing of this book, I have benefitted greatly from my affiliation with Cornell University, particularly the Southeast Asia Program and Mann, Olin and Kroch Libraries. The Southeast Asia Program represented a fount of expertise and comparative materials. They contributed financially to my March 2019 field trip to East Kalimantan, and the libraries granted me easy access to abundant literature and resources and to my own fieldnotes, some of which I had already archived there.

I am also grateful to the Center for International Forestry Research in Bogor, Indonesia, which has also supported my work in a variety of ways. The March 2019 fieldwork was partially supported by a PIM grant (administered through the Consultative Group of International Agricultural Research's gender platform). My graphics were beautified and funded by the Forests, Trees and Agroforestry Program. I particularly thank Anne Larson, Markus Ihalainen, Iliana Monterroso and Rahayu Koesnadi at CIFOR and a previous CIFOR scientist, Bimbika Sijapati Basnett, for their support at various stages throughout this process. I also thank Gideon Suharyanto, Sufiet Erlita and Dharmi Bradley for significant help with visuals and some statistics. Thanks also to CIFOR and Biodiversity (specifically Marlène Elias), for their contribution to making this book open access.

Various parts of the manuscript benefitted greatly by the critical and constructive reviews of Tim Babcock, A. Michael Colfer, Michael Cummings, Rebecca Elmhirst, Adi Gangga, Chris Gibson, Emily Hillenbrand, Iliana Monterroso, Susan Paulson, Gwendolyn Pierce, Jack Putz, Maureen Reed and Jeff Sayer, as well as two anonymous reviewers (some of whom read the entire manuscript more than once)—though none of these individuals bears any responsibility for any remaining shortcomings.

And finally I thank the editors at Earthscan (Hannah Ferguson, John Baddeley) and Ramachandran Vijayaragavan, Project Manager) who were enthusiastic, responsive, and helpful throughout the publication process.

GLOSSARY

Adat Custom, Indonesian

Agency Capacity of individuals to act independently and to make their own free choices

Amenity migration "In-migration that occurs in a place because people are drawn to its natural and social features" (Charnley, McLain, and Donoghue 2008, p. 744)

Ball buster A dominating (or assertive) woman seen as having little regard for men's sensitivities related to their masculinity (informal)

Bilaterality Custom of tracing descent through both fathers and mothers

Bupati District or regency head, Indonesian

Camat Sub-district or county head, Indonesian

Companionate marriage Marriage that stresses affection, comradeship, democracy and happiness of members of the family (Bernard 1982)

Extractive research Data taken from communities with little or no feedback of information or direct benefit to them

Gender "Gender [is] a sociocultural system that organizes the practices and relationships that play out among humans, and between humans and their environment, infusing them with power and meaning that refers symbolically to sex and sexuality" (Paulson 2016, pp. 1–2)

Hegemonic 'Dominant' or 'hegemonic' representations or ideologies: "those models that support the claims of a particular category of people to superior status and power, models which are most likely to be invoked in formal discourse and which are most often accorded a position of supremacy among other, potentially competing models" (Brenner 1995, p. 21)

Hegemonic masculinity Hegemonic masculinity refers to practices that legitimize men's dominant position in society and justify subordination of women and of men with non-dominant qualities and practices. Initially in the masculinities literature it was discussed as a culturally idealized form of manhood that was socially and hierarchically exclusive and concerned with breadwinning, was anxiety-provoking and differentiated (internally, hierarchically), brutal and violent, pseudo-natural and tough, psychologically contradictory and thus crisis-prone, economically rich and socially sustained, and applied to all men (Donaldson 1993).

Heteronormativity The view that heterosexuality, attraction to the opposite sex, is the only natural, acceptable sexuality

Iban People originally from the island of Borneo (here, in West Kalimantan), typically speaking their own language

Identity How a person self-identifies, how he thinks of himself, what elements of his being he values and what differentiates him from others (also applicable to women)

Javanese People originally from central and eastern parts of the island of Java, who typically speak the language of that region

Jorong Hamlet, Indonesian

Kabupaten District or regency, Indonesian

Kecamatan Sub-district or county, Indonesian

Kenagarian Minang administrative unit above a village, Minang

Kenyah People originally from the island of Borneo (most in East Kalimantan), typically speaking their own language

LGBTQ Lesbian Gay Bisexual Transsexual and Queer

Local Here, a social structural division in Bushler Bay; families in which the husband worked in private industry and typically had close familial ties in the region

Matrilineality Custom of tracing descent through the mother's line (mother to daughter to granddaughter)

Matrilocal Custom whereby a man moves in with his wife's family on marriage

Matrifocal Kinship, social relations and structures focused on the mother or on women more generally

Matrilateral Kin relations on the mother's side or through the mother

Melayu General term that can refer to many Indonesian ethnic groups, but which also refers to a specific group in West Kalimantan

Minang Short form of Minangkabau (West Sumatra's dominant ethnic group)

Minangkabau People originally from the province of West Sumatra, who typically speak the language of that region

NGO Non-governmental organization

Normative Designation of some actions or outcomes as good, desirable, permissible; others as bad, undesirable, not allowed

NTFP Non-timber forest product

Patrilineality Custom of tracing descent through the father's line (father to son to grandson)

Patrilocal Custom whereby a woman moves in with her husband's family at marriage (also termed virilocal)

Patrifocal Kinship, social relations and structures focused on the father or on men more generally

Pecker Penis (informal)

Performative View of actions, behaviours and gestures as both the result of an individual's identity and contributing to one's identity formation. Identity is continuously being redefined through speech acts and symbolic communication.

Peter Penis (informal)

Public Employee Social structural division in Bushler Bay, referring to families in which the husband worked for a public institution

Pussy whupped Under one's wife's domination, unable to resist her sexuality and exert manly control (informal)

Sundanese People originally from the western part of the island of Java, who typically speak the language of that region

Transmigration Longstanding Indonesian programme to move people from densely populated Java and Bali to the 'Outer Islands' of Indonesia

References

Bernard, Jesse. 1982. *The Future of Marriage (1982 edition)*. New Haven, CT: Yale University Press.

Brenner, Suzanne A. 1995. "Why Women Rule the Roost: Rethinking Javanese Ideologies of Gender and Self-Control." In *Bewitching Women, Pious Men: Gender and Body Politics in Southeast Asia*, edited by Aihwa Ong and Michael G. Peletz, 19–50. Berkeley: University of California Press.

Charnley, Susan, Rebecca J. McLain, and Ellen M. Donoghue. 2008. "Forest Management Policy, Amenity Migration, and Community Well-Being in the American West: Reflections from the Northwest Forest Plan." *Human Ecology* 36:743–761. https://doi.org/10.1007/s10745-008-9192-3.

Donaldson, Mike. 1993. "What Is Hegemonic Masculinity?" *Theory and Society* 22 (5):643–657.

Paulson, Susan. 2016. *Masculinities and Femininities in Latin America's Uneven Development*, 1–2. New York and London: Routledge.

FOREWORD

Susan Paulson

Multiple masculinities shed new light on forest management

This adventurous book takes readers on a journey through the author's long career in international conservation and development, exploring findings about diverse forest communities together with personal positions and experiences. A wealth of tangible detail gathered along the way allows readers to see and hear how diverse people-forest relationships are informed by gendered identities, knowledges and opportunities.

Through decades of research and practice, Colfer has observed that fellow scientists perceive forests as masculine realms. Yet lack of attention to the variability of masculine identities, norms and behaviours has resulted in incomplete data and analysis, and has limited efforts to tailor studies and projects to fit actual forest realities.

Masculinities in Forests supports more comprehensive research and practice by demonstrating mutual influences between masculine identities and environmental management, and by applying concepts of plural masculinities and intersectionality to varying manifestations of those processes. It challenges universalizing stereotypes about men by presenting unique empirical findings, and by showing how they vary across sociocultural, ethnoracial and ecological contexts. This book will support foresters, ecologists, natural resource managers and gender scholars to strengthen their attention to men and masculine identities, thereby increasing the rigour of empirical research and enhancing the design and outcomes of policies and projects.

Colfer's work resonates with my own journey exploring processes through which gender, class and ethnoracial systems interact with farms and forests. Like Colfer, I was raised in the US, and have worked in wide-ranging contexts which, in my case, included 15 years living full-time in South America and six years in Europe. My research in Andean and Amazonian communities has long been enriched by Colfer's writing and edited collections (Colfer 2005; Colfer, Basnett, and Elias 2016; Colfer et al. 2017), and I have enjoyed reading Colfer's work with students pursuing graduate degrees in natural and social sciences. In a course I am teaching now on masculinities and environment, participants are especially inspired by *Making Sense of 'Intersectionality': A Manual for Lovers of People and Forests* (Colfer, Basnett, and Ihalainen 2018). I am honoured to introduce this book, and do so by drawing ideas and phrases from strands of my own intellectual trajectory that have been influenced by Colfer's work, including two of my books, *Masculinidades en movimiento. Transformación territorial* (2013) and *Masculinities and Femininities in Latin America's Uneven Development* (2015) and other publications (Paulson 2016; Paulson and Boose 2019).

Path-breaking contribution to complex challenges

Rachel Carson's *Silent Spring* (1962), Carolyn Merchant's *The Death of Nature: Women, Ecology and the Scientific Revolution* (1980) and others brought attention to perspectives and behaviours that are linked to masculinity in western culture, and that have impelled increasing domination and exploitation of natural environments. In light of many ensuing portrayals of masculinity as toxic, and of men as degraders of women and nature, it is no surprise that few men wanted to hear any more about gender and environment.

In a welcome shift, Colfer's cross-cultural material and intersectional analyses take us beyond dominant stereotypes of destructive masculinity to reveal surprisingly varied practices and meanings; vital among these are different forms of knowledge, nurture and love for forests.

Current scholarship conceptualizes plural masculinities as constellations of qualities, behaviours, attitudes and accomplishments that are associated with the category "man" in particular communities of interpretation, and that shift through time and across contexts (Hultman and Pulé 2018; Connell 2017). Logging, for example, has been ranked world-wide among the most masculine, and the most dangerous, fields of work. Yet the forestry-manliness nexus is far from static or homogenous. Loomis (2017) reveals historical forces behind shifting ideas of masculinity and nature among working-class loggers in the US Pacific Northwest, while Brandth and Haugen (2000) trace shifts in media representations of Norwegian forest workers from lumberjacks to business managers.

Here, in one volume, we get to know different men engaging forests in different ways: career loggers, professional forest managers, swidden farmers, NGO practitioners, hunters, gatherers and a group that usually escapes the lens of gender analysis: research scientists and international development experts. In Chapter 4, for example, Colfer shows how a particular mosaic of primary and secondary forest with new and older swiddens provides men in Long Segar opportunities to develop and demonstrate forest-related skills including hunting, fishing, gathering forest products and finding their way through dense forest.

Why should you read this book?

This book will motivate and empower you to recognize gender realities unique to the places where you live and work. Attention to masculine actors and phenomena that are most certainly relevant for forest management has been constrained by the practice of lumping 3.8 billion humans into one category labelled with the veterinary term "male," and by the assumption that all members of said category enjoy privilege and power, and do not need or want gender-based attention. This book demonstrates how to use empirical evidence and intersectional analysis to see beyond those stereotyped categories.

One powerful way to transcend gross generalizations is by paying attention to descriptions of observed practices, characteristics and voices of differently positioned actors, including loggers, farmers, professionals and entrepreneurs. The evidence-based case studies presented here help us to see how masculinities interact with communities and ecosystems in half a dozen contexts:

- The author's own family history and upbringing in various US communities
- Loggers and public employees in the forests of the American Pacific Northwest in the mid-1970s
- Uma' Jalan Kenyah Dayaks in the wilds of Borneo in the early 1980s
- Javanese, Sundanese and Minangkabau farmers in the more domesticated forests of West Sumatra in the mid-1980s
- Global professionals in international conservation and development from the 1980s to the 2010s
- 21st century return visits to earlier sites

Another useful way to see beyond stereotyped categories is by paying attention to race, class and other social systems that shape different groups of men and women. This book describes how men's relationships with forests are influenced by Christian and Muslim traditions, by matrilineal and bilateral kinship and by class and vocational positioning. In doing so, it masterfully demonstrates the approach called intersectionality, which has been applied widely to make visible conditions of women whose identities are influenced by multiple kinds of difference. Colfer widens the scope of intersectional analysis to encompass diversely positioned men residents, and also to consider professionals who study, make and implement policies and programmes. Moving beyond assumptions that all men are dominant oppressors who control resources and monopolize power over women, this approach encourages readers to recognize that masculine positions of some forest-dwellers are co-constituted by racial, occupational or other identities that disadvantage them vis-à-vis some other men and women. It also entails recognition that, in each context, certain women (like scientist Dr. Carol J. Pierce Colfer in Indonesia) enjoy class, professional and/or ethnoracial positionality tied to greater privilege and power than that accessed by some men.

Reading this book will help you to understand gender systems by looking at mutually-influencing interactions among diverse masculinities, femininities and other identities, just as you understand ecological systems by looking at mutually-influencing interactions among diverse plants, animals and other elements.

For half a century, feminist science and gender studies have challenged dominant interpretations of environmental issues and motivated new kinds of action. Early work focusing on women and women's issues forged vital advances—notably revealing that not only men, but also diverse women, engage actively in forest management around the world. But some women-focused programmes have provoked negative reactions and backlash. Some men have perceived women's empowerment as a zero-sum game that entails dis-empowering men. Changes in women's conditions have—in some contexts—been blamed for a range of social and ecological problems. Rather than lose faith in the benefits of paying attention to gender, Colfer advances the struggle to develop more comprehensive approaches.

Over time, feminist research and action have contributed to emerging considerations of men and masculinities. Increased awareness of ways in which hierarchical gender institutions constrain and harm men (as well as women and others) began motivating some men to question, resist and adapt gender norms and practices that they experience as degrading or dangerous to themselves and others, including their natural environments. New work on masculinities and forests began to emerge as a fruit of, and a vital complement to, decades of women-centred research and projects on gendered knowledge and participation in forest management (e.g., Agarwal 2009, 2010; Colfer, Basnett, and Elias 2016; Colfer et al. 2017; Nightingale 2003).

Even readers committed to sustaining a focus on women will benefit from a more comprehensive framework to understand and address the sociocultural systems that produce and reproduce gender arrangements that variously constrain and oppress women, or support their healthy development. More inclusive and systemic approaches to gender often lead to greater interest and participation by diverse members of communities and professional teams.

Finally, you should read this book to strengthen responses to ongoing changes and socio-ecological crises. As this book is prepared for publication, in April 2020, life around the world is undergoing dramatic changes amid COVID-19, a pandemic that has much to do with masculinities and with forests. To date, many more men have died from the coronavirus than women, in some countries nearly twice as many. This must be studied in relation to factors already causing men to suffer disproportionate rates of premature death, leading to gender gaps by which women, on average, outlive men in nearly every country, gaps that have been increasing rapidly in most countries. Just as forest management has been a factor in wildfires, floods and other intensifying disasters, so too has it influenced recent epidemics (HIV/AIDS, MERS, SARS and Ebola). The growing ease with which viruses jump between animal species, and from other animal hosts to humans, is conditioned by expansion of corporate agricultural systems, including tree plantations; by encroachment of humans on forest habitats; and by the commodification of forest wildlife. There is no doubt that this pandemic, together with other socio-natural disasters, will impact future human-forest relations.

Gender-specific documentation of past processes of change in environmental management will be helpful in anticipating future changes, and in developing more effective and resilient strategies for managing forests and collaborating with those who live in and near them. The temporal span of Colfer's research shows that gender-forest relations change in response to historical forces and factors, and also to human visions and motivations. The generational trajectories of Colfer's grandparents, parents, herself and her children provide compelling evidence of our ability to visualize, negotiate and change gender identities and relations.

In each of the realms studied, Colfer identifies some masculine-identified behaviours and attitudes that are intertwined with processes that are risky and destructive to human and environmental well-being, and others that support, and can further support, the well-being of humans and non-human nature. She makes clear that, in order to see and support positive features associated with masculinity in each context, forest management programmes will have to move away from unidirectional extension service. Methods built on mutual respect and two-way communication will allow employees of national park and forest services, the Forest Research Institute and other organizations to learn about local forest-related knowledge, goals and problems, and to recognize and work with strengths of diverse men and women in the face of changes and challenges.

This book on forest management also sheds light on other ventures that threaten forests and exacerbate climate change: mining, petroleum exploitation, ranching and agroindustry. Like logging, these sectors are characterized by exceptionally high ratios of men to women workers, and by extraordinarily high rates of occupational accidents, illness and death. However, simply characterizing these industries as masculine is insufficient to understand their gender dynamics, and inadequate to respond to troubles. By inspiring research on masculinities in each of these parallel realms, the present book can make a wider impact on efforts to address ecosystem and earth system challenges.

Conservation and development processes better suited to diverse empirical realities

Colfer's boldest move is to turn a gender lens on her own positionality as researcher and conservation professional, and on the colleagues and organizations with which she has collaborated. This self-analysis complements Colfer's research on forest communities to help address tendencies that continue to undermine sustainable development efforts: the exclusion of men from gender-related analysis and support, and the categorization of all "men" as a homogenous group.

At the turn of the century, the possibility of incorporating men into gender and development policies and projects was raised and debated heatedly (e.g., Cornwall and White 2000). Chant and Gutmann (2002, p. 269) observed that

> Gender and Development (GAD) policies encompass a broad range of approaches and interventions, but to date have largely been associated with programmes established by women for women. This is despite the fact that, in theoretical terms, GAD is concerned with gender relations, and therefore with men as well as women.

During two decades following these calls, a number of programmes have addressed specific groups of men and conditions of masculinity, and even supported third gender identities. Most policy-makers and programme-planners, however, have resisted doing so (Lind 2010).

In the meantime, scholarship on masculinities began to move beyond static man-woman binaries, to pay attention to intersectionality and context and to reveal how specific practices and representations of masculinity gird power structures that drive economic and environmental degradation or sustainability (Bannon and Correia 2006; Cornwall, Edström, and Greig 2011; Parpart 2015). Research on case studies has identified masculine norms that interfere with projects encouraging reforestation (Gonda 2017), as well as masculine commitments to nurture biodiversity and healthy ecosystems that can support conservation programmes (Bolt Gonzáles 2003; Devore, Hirsch, and Paulson 2020). Will Boose and I were happy to document these emerging trends in research and scholarship in a review of 160 recent publications on masculinities and environment (Paulson and Boose 2019).

Colfer's book has unique potential to bridge this troubling divide between practice and theory by communicating scholarly understandings to scientists and professionals who implement projects around the world. Among the deepest conceptual challenges faced by these professionals is that of binary thinking, manifest in the habit of perceiving and dividing the world's diverse residents into two presumably innate categories.

Scholars have long agreed that gender identities are not constrained by biological dimorphism. Ethnographic, historic and archaeological evidence reveals widely diverse arrangements in which community members live as more than two gender identities. Across South Asia, multiple identity formations, frequently grouped together under the term "hijra" have existed for centuries, and still today, in numbers surpassing a million (Goel 2019). After repressing these gender expressions during colonial and national periods, countries including India, Pakistan, Bangladesh, Nepal and Thailand, now recognize a third gender category on national identification cards, driver's licenses and other official forms. Closer to some of Colfer's field sites, Graham-Davies (2004) describes five gender categories in South Sulawesi, Indonesia, while Keeler (2017) investigates complementary masculine paths in Burma: that of Buddhist monk and head-of-household.

In recent centuries, colonization, modern development and globalization have disseminated polarized man-woman categories in many parts of the world, creating the false appearance that these categories express an innate and universal structure of human life. At the time of European contact, however, over a hundred instances of diverse gender expression were documented among Native American tribes (National Park Service 2018). Although violent efforts were made to annihilate forms perceived to vary from the European binary, some centuries-old traditions continue today in identities referred to as Muxes, two-spirit people and numerous native terms (Werft and Sanchez 2016). Similarly, Colfer's studies show that, in spite of the institutionalization of hierarchical man-woman distinctions by churches, mosques, government and industries, community members in Long Segar have maintained remarkably muted expressions of man-woman binary.

While engaging established conversations governed by prevailing categories (man vs. woman), Colfer's book also presents perspectives and materials that critically challenge aspects of those conversations. Qualitative findings brought together here support more culturally aware ways of interpreting gender patterns observed in the field, which can contribute to the ongoing development of more appropriate—and more robust—quantitative methods and instruments. In short, the rigour, empirical specificity and practical usefulness of all types of research and professional practice are strengthened by recognizing gender identities that exist—even when there are more than two of them.

There is no doubt that the use of common categories (man vs. woman) facilitates understanding and dialogue with a wide range of interlocutors, and enables the incorporation of gender language into national and international policy and legislation. At the same time, this book shows that the universal category "man" comes up short in efforts to represent empirical realities constituted by different norms and expressions of masculinity, linked with different positions of power, knowledge and decision. It is vital to recognize differences among the masculinity of entrepreneurs investing in forest concessions, that of loggers hired to clear trees and that of indigenous dwellers in the forest in question. Binary gender language also limits efforts on the political and strategic front: first, by reinforcing ideological messages that gender roles are determined by sex difference; and second, by making it difficult for many to think beyond a framework of competition or antagonism between two teams—men versus women.

Colfer's careful empirical documentation of diverse masculinities will motivate and empower efforts to change gender-environment arrangements for the better by demonstrating that gender roles and expectations are not shaped by sex biology, but by historical processes, which change and can be changed. As Connell (2017, p. 5) writes, "It is true that many men and influential forms of masculinity are involved in environmental destruction. But not because XY chromosomes mechanically generate bad behavior." What needs to be redirected is not human nature, but the historical production of masculinities, femininities and gendered divisions of labour and power that have driven and enabled degradation of forests and earth systems.

Conclusion

Countless studies and projects have striven to improve conservation and forestry programmes by enhancing recognition and incorporation of diverse women.

That is still a valuable goal and work in progress. To date, most environment-related efforts are also hindered by gender stereotyping of men and by lack of attention to gendered conditions, identities and expectations associated with diverse masculinities. This evasion has taken a toll scientifically, resulting in unbalanced and incomplete empirical records, and politically, contributing to resistance among researchers, practitioners and activists who are unwilling or unable to incorporate boys and men into gender-focused work.

This book arrives at an extraordinarily fertile moment for improving conservation and forestry with thought and action around masculinities. Unprecedented changes and challenges in global human-environment relations are generating tensions and opening possibilities for innovation among diverse expressions of masculinity (Gaard 2014; Hultman and Pulé 2018; Messerschmidt and Messner 2018). By helping us to understand and support these dynamics, the present book contributes to enhancing the viability and resilience of forests and broader environments, together with human populations who reside in and near those forests.

Susan Paulson
University of Florida
April 2020

Sources

Agarwal, Bina. 2009. "Gender and Forest Conservation: The Impact of Women's Participation in Community Forest Governance." *Ecological Economics* 68:2785–2799. https://doi.org/10.1016/j.ecolecon.2009.04.025.

Agarwal, Bina. 2010. *Gender and Green Governance: The Political Economy of Women's Presence Within and Beyond Community Forestry*. Oxford: Oxford University Press.

Bannon, Ian, and Maria Correia. 2006. *The Other Half of Gender: Men's Issues in Development*. Washington, DC: World Bank.

Bolt Gonzáles, Alan. 2003. *Masculinidades y Desarrollo Rural: Una Nueva Manera de Satisfacer las Necesidades Humanas Esenciales y Defender la Red de Vida (GAIA)*. Managua, Nicaragua: SIMAS.

Brandth, Berit, and Marit Haugen. 2000. "From Lumberjack to Business Manager: Masculinity in the Norwegian Forestry Press." *Journal of Rural Studies* 16 (3):343–355.

Chant, Sylvia, and Matthew Gutmann. 2002. "'Men-streaming' Gender? Questions for Gender and Development Policy in the Twenty-First Century." *Progress in Development Studies* 2 (4): 269–282.

Colfer, Carol J. Pierce. 2005. *The Equitable Forest: Diversity, Community and Resource Management*. Washington, DC: RFF/CIFOR.

Colfer, Carol J. Pierce, Bimbika Sijapati Basnett, and Marléne Elias. 2016. *Gender and Forestry: Climate Change, Tenure, Value Chains and Emerging Issues*. London: Earthscan/Routledge.

Colfer, Carol J. Pierce, Bimbika Sijapati Basnett, and Markus Ihalainen. 2018. "Making Sense of 'Intersectionality': A Manual for Lovers of People and Forests." *CIFOR Occasional Paper* 184: 40.

Colfer, Carol J. Pierce, Marléne Elias, Bimbika Sijapati Basnett, and Susan Stevens Hummel. 2017. *The Earthscan Reader on Gender and Forests*. London: Earthscan/Routledge.

Connell, R. 2017. "Masculinities in the Sociocene." In *Men and Nature: Hegemonic Masculinities and Environmental Change*, edited by S. MacGregor, and N. Seymour, 5–8. Munich: RCC Perspectives.

Cornwall, Andrea, Jerker Edström, and Alan Greig. 2011. *Men and Development: Politicising Masculinities.* Chicago: University of Chicago Press.

Cornwall, Andrea, and Sarah C. White. 2000. "Men, Masculinities and Development." *Institute of Development Studies* 31 (2):1–6. https://doi.org/10.1111/j.1759-5436.2000.mp31002001.x.

Devore, Jonathan, Eric Hirsch, and Susan Paulson. 2020. "La conservation nourrissant avec la nature humaine et autre nature au Brésil." *Anthropologie et Sociétés.* Numéro 43–43.

Gaard, Greta. 2014. "Towards New Ecomasculinities, Ecogenders, and Ecosexualities." In *Ecofeminism: Feminist Intersections with Other Animals and the Earth*, edited by Carol J. Adams and Lori Gruen, 225–240. New York: Bloomsbury.

Goel, Ina. 2019. "India's Third Gender Rises Again." *Sapiens.* www.sapiens.org/body/hijra-india-third-gender/.

Gonda, Noémi. 2017. "Rural Masculinities in Tension: Barriers to Climate Change Adaptation in Nicaragua." *Environment and Society* 4:69–76. https://doi.org/10.5282/rcc/7985

Graham-Davies, Sharyn. 2004. "Hunters, Wedding Mothers, and Androgynous Priests: Conceptualising Gender among Bugis in South Sulawesi, Indonesia." PhD dissertation, Anthropology and Asian Studies, University of Western Australia.

Hultman, Martin, and Paul Pulé. 2018. *Ecological Masculinities: Theoretical Foundations and Practical Guidance.* Routledge Studies in Gender and Environments. New York: Routledge.

Keeler, Ward. 2017. *The Traffic in Hierarchy: Masculinity and Its Others in Buddhist Burma.* Honolulu, HI: University of Hawai'i Press.

Lind, Amy. 2010. *Development, Sexual Rights and Global Governance.* London and New York: Routledge.

Loomis, Erik. 2017. "Masculinity, Work, and the Industrial Forest in the US Pacific Northwest." *RCC Perspectives: Transformations in Environment and Society* 4:37–43. https://doi.org/10.5282/rcc/7981.

Messerschmidt, James W., and Michael A. Messner. 2018. "Hegemonic, Nonhegemonic, and 'New' Masculinities." In *Gender Reckonings: New Social Theory and Research*, edited by James W. Messerschmidt, Patricia Yancy Martin, Michael A. Messner, and Raewyn Connell, 35–56. New York: New York University Press.

National Park Service. 2018. "Gender and Sexuality in Native America: Many People, Many Meanings." *United States National Park Service* (NPS). www.nps.gov/articles/gender-and-sexuality-in-native-america.htm.

Nightingale, Andrea. 2003. "A Feminist in the Forest: Situated Knowledges and Mixing Methods in Natural Resource Management." *ACME: An International E-Journal for Critical Geographies* 2 (1):77–90.

Parpart, Jane L. 2015. "Men, Masculinities, and Development." In *The Routledge Handbook of Gender and Development*, edited by Anne Coles, Leslie Gray, and Janet Momsen, 14–23. New York: Routledge.

Paulson, Susan. 2013. *Masculinidades en Movimiento. Transformación Territorial y Sistemas de Género.* Buenos Aires: TESEO.

Paulson, Susan. 2015. *Masculinities and Feminities in Latin America's Uneven Development.* New York: Routledge.

Paulson, Susan. 2016. "Toward a Broader Scope and More Critical Frame for Intersectional Analysis." In *Palgrave Handbook of Gender and Development: Critical Engagements in Feminist Theory and Practice*, edited by Wendy Harcourt, 395–412. London: Palgrave.

Paulson, Susan, and Will Boose. 2019. "Masculinities in Environment." *CAB Reviews* 14 (30). https://doi.org/10.1097/00007435-200101000-00004.

Werft, Meghan, and Erica Sanchez. 2016. "Male, Female, and Muxes: Places Where a Third Gender Is Accepted." *Global Citizen*. www.globalcitizen.org/en/content/third-gender-gay-rights-equality/.

1

MASCULINITIES IN FORESTS

Here and there, then and now

Born in suburban Chicago, raised in the cornfields of the Midwest and the arid plateau of Ankara, Turkey, still I have loved forests since I first encountered them. Beautiful images, tinged with the frisson that accompanies real adventure, come to mind as I consider my topic here:

Driving up the eastern shore of the Olympic Peninsula in the North American State of Washington. To my right is Hood Canal, an offshoot of salty Puget Sound; to my left are the magnificent snow-covered Olympic Mountains, their foothills covered in tall and stately evergreen forests—hemlock, pine, cedar, Doug fir, small streams breaking any possible monotony, herds of elk browsing in the area cleared for power lines.

Later in Borneo, the weather is warm and the trees are again evergreen, but different species: *Dipterocarpus*, *Dryobalanops* and *Shorea*, majestic species like *Koompassia excelsa*, and iron-hard ones like *Eusideroxylan zwageri*. I sit in a canoe, navigating a narrow freshwater stream, foliage bending over us, creating a green tunnel through which we manoeuvre. The paddles gently break the water, gibbons call to each other in the early morning hours.

On another journey further inland, I'm taken by canoe to report to local officials. The young Kenyah men kindly and energetically pole me upstream, over fast and noisy rapids, their bare muscles bulging in the sunlight, straining against the water's power with all their might. We pass various patches of forest—old growth, young and old secondary regrowth and swidden rice fields here and there.

Yet again, I go with a log truck driver into the forest to watch American logging in the 1970s. The logger drives the huge log truck up dirt roads, through dense forest, past clear cuts and replanted areas with trees of varying ages and heights. I am entranced by the habitat[1] and intrigued by these men who value their own strength and their outdoor profession so highly. As a feminist, I am part repelled by the inherent sexism, part attracted. As others have noted, these men are sexy in their 'in-your-face' masculinity:

> Let me tell you about loggers. I married one 18 years ago. I couldn't resist the smell of utter maleness in damp, sweaty black wool underwear sweetened by the heavy scent of fresh sawdust and chainsaw exhaust . . . I still love the romance of big men, big machines and big trees . . . Loggers are the last of a dying breed of men who "work" for a living. Their work is dangerous and back breaking . . . [An environmentalist] should try hugging a logger—she'll never go back to trees!
>
> *(Maureen Henderson,* Chilliwack Times, *1993, p 9)*[2]

Here, I hope to take you on a journey to five 'spaces':

- The America of my childhood, from Maine through the Midwest to the Pacific Northwest
- The forests of America's Pacific Northwest in the mid-1970s
- The wilds of Borneo in the early 1980s
- The somewhat more domesticated forests of West Sumatra in the mid-1980s, and
- The world of international conservation and development from the 1980s to the 2010s.

I then add an update on the rural US and Borneo more recently to see how these worlds of men and forests have changed.

My hope is that by visiting these spaces, we can see first how masculinities interact with forests and forest management, and how forest managers can manage forests better based on such knowledge. Secondly, I hope that my conceptualization of cultural stability and men's agency, in interaction, can cast light on ordinary men's lives in their various contexts—contributing also to gender studies.

In my professional life, I paid a lot of attention to women. But a few years back, I began to feel a sense of malaise. We said we were looking at gender,[3] whereas in reality mostly we were looking at women's worlds. I began to wonder about men's worlds. As I read the very different body of literature called 'masculinity studies' the seeds for this book were planted. The ideas took hold and grew.

My desire to share my findings with gender and masculinity scholars on the one hand and biophysical scientists (foresters, ecologists, natural resource managers) on the other, creates problems (epistemological, evidentiary and in terms of vocabulary). I address this problem by providing warning, in footnotes if not clear from subtitles, regarding sections of greater or lesser interest to one or another group of readers.

One impetus to writing this book was my initially unreasoned resistance to the phrase 'hegemonic masculinity'—once so common in masculinity studies.[4]

Although I recognized that men have advantages related to power, apparently in all cultures, the rationales for such advantages were too variable to be considered globally 'hegemonic', and I wasn't sure one could always identify one version even as locally hegemonic everywhere either.[5] Men, their behaviour and the norms that influenced them, varied enormously from place to place, and even man to man. Over time, I also realized that writings about hegemonic masculinity were shifting to a focus on *masculinities*, plural. I began to interrogate my ethnographic experience, much of which has been in forests, settings typically considered masculine. Forests are the domain of logging and timber production, heavy equipment and international markets, and the field of forestry has been dominated by such stereotypically male concerns. Although much of my professional life has been devoted to demonstrating women's presence, interests and activities in forests, I realized that I also knew quite a bit about what men do in forests and among the forest peoples I know well.

I'd also read books and articles on masculinity over the years and came away with the disappointing feeling that this literature focused almost exclusively on *problems* with masculinity.[6] There was an inordinate amount written on domestic abuse, HIV/AIDS, warfare and homosexuality (the last no longer seen as problematic itself, though often spelling problems for those so inclined).

In late 2017, I began to dip methodically into such studies. I found more varied literature that dealt with a wider spectrum of men's lives than I had hitherto seen. There remained a clear urban bias though in many studies, as well as an emphasis on nonconventional sexuality. I knew from my research and life experience that men's lives were not composed only of problems, that men—ordinary heterosexual men—had strengths and contributions to make in family and community life.[7] I also had seen the incredible variety in ideals of manhood, in its practice and in men's goals and interests. This book represents an attempt toward 'righting the balance' and sharing what I know of the lives of predominantly heterosexual men in rural, particularly forested, places.

Little (2006) and Pini, Brandth and Little (2015) point out the paucity of rural studies generally, an observation even more cogent for the topic of masculinities in forests. In this book, I reflect on the variety of masculinities I encountered, primarily in three rural, forested contexts—all areas where I lived, doing ethnographic research over long periods of time. I then turn to the international research world, from which forest-related research is routinely conducted, examining notions of masculinity among elites.

Much of the masculinity literature (with a few notable exceptions) is written by men about men, just as gender materials on women are often written by women. In trying to achieve a more balanced view of men's and women's lives, I look here at men through an American woman's eyes. As in Enria's (2016) work, masculinity has not been the primary focus of my research. What I present here represents a mining of my ethnographic experience, which inevitably included both men's and women's beliefs, behaviour and goals. This 'excavation' uncovered useful understandings of men's lives.

Keeler (2017) warns us that the assumptions that westerners, devoted to egalitarian ideals, hold can prevent them (us) from seeing hierarchies assumed by others (in his case, the Burmese). Similarly, as a woman with what might be a 'womanly' world view and set of assumptions, I may be blind to assumptions men hold; I cannot know what is in the hearts of men. Yet I can observe and see the *effects* of what is in their hearts. And I can make educated guesses about what such actions imply—as I do when trying to understand and convey the meanings and actions of the Kenyah or the Minangkabau in Indonesia or a to-me-once-alien American subculture.

Partly because of these uncertainties, I present an account of these ethnographic experiences with a reflexive bent, one that both recognizes my own positionality and considers the effects of the passage of time, my own aging process and likely influences on my observations and interpretations.[8] I approach this task by taking the reader on a journey through space and time—linking my own life journey with the passage of time and travel through space.

Masculinities and forests

But what does all this have to do with forests? Most fundamentally, the general agreement that forests are 'masculine contexts', at least in the international world of conservation and development, suggests that we might want to know more about this defining quality. What are the implications of masculinities in forests? How do they play out in different contexts? How do men's relations with forests differ from women's?

I agree with Paulson (2016), who has encouraged us to look more carefully at the various masculinities extant in particular geographic spaces. She says,

> The processes of becoming, and the practices of . . . [a particular identity].[9] . . . depend not only on intersubjective relations with people in similar and different positions within a social system but also on relations in and with the biophysical space that has been shaped by, and that works to reproduce, those social systems.
>
> *(p. 153)*

Forests are one such biophysical space.

In the coming chapters, I look at the elements of masculinities that men in these different contexts, times and ages choose from among the structured options available to them. The emphasis in many, though not all, contexts on men-as-providers suggests implications for forest-based activities. Men may be more likely to view forests as potential sources of income than women (as some studies have shown). Some men may value the forest primarily for its products (e.g., Sitiung, Chapter 5). The devotion to an outdoor life as an indicator of manliness may give forests special meaning to men in some communities (e.g., Bushler Bay loggers, Chapter 3). The desire to demonstrate physical strength may be a factor in some men's desires to remain in logging, even when formal management goals have shifted to another use. The imperative to demonstrate courage can be another motivator to work in the woods (e.g., Chapters 3 and 4, on Borneo's Kenyah). Emphasizing control and articulateness, as some elite men do (Chapter 6), can have adverse implications for their ability to hear what local communities desire and know.

Men's ideas about their own identities and related practices are important to forest management insofar as the forestry world is committed to taking human beings into account—a commitment that has only grown in recent years. There have been increasing attempts, for instance, to involve local communities in collaborative management (building on works such as CIFOR's Adaptive Collaborative Management Program, e.g., Colfer 2005; or more recently, in relation to REDD+, Larson et al. 2016). Extractive studies[10] of forest use are also common, but generally without attention to people's ideas about what a good man is or should be (cf. Petesch et al. 2018; Elias et al. 2018, for counter-examples)—something that can influence what men want, need and are willing to do in forests.

Forest policy development could also benefit from a better understanding of the varieties of ways men interact with and value (or disvalue) forests. If local people link men's masculinities primarily to provisioning from the forest as a source of products (and women do not have other important uses), the development of economic alternatives like agriculture as the forest disappears may be a good policy option. However, if local men's identities are tightly bound up with the forest[11]—whether to protect it, recreate in it, or harvest it—then a policy that switches management from timber to oil palm, for instance, will cause more personal and cultural dislocation (Colfer 2018). Similarly, decisions to resettle men who have strong links to forests to an agricultural or urban landscape will engender a serious sense of loss. For those with fewer direct forest-culture connections, the pain associated with such a move may be far less.

Besides the value for foresters, these studies contribute to our social scientific understanding of gender, something we (gender specialists) have approached in a one-sided manner, often examining only women's lives, as noted earlier.

What kinds of forests?

The term 'forest' can be interpreted in a variety of ways, including even as a previously forested area. A commonly used definition is that of the Food and Agriculture Organization (FAO):

> Forest includes natural forests and forest plantations. It is used to refer to land with a tree canopy cover of more than 10 percent and area of more than 0.5 ha. Forests are determined both by the presence of trees and the absence of other predominant land uses. The trees should be able to reach a minimum height of 5 m. Young stands that have not yet but are expected to reach a crown density of 10 percent and tree height of 5 m are included

> under forest, as are temporarily unstocked areas. The term includes forests used for purposes of production, protection, multiple-use or conservation (i.e. forest in national parks, nature reserves and other protected areas), as well as forest stands on agricultural lands (e.g., windbreaks and shelterbelts of trees with a width of more than 20 m), and rubberwood plantations and cork oak stands. The term specifically excludes stands of trees established primarily for agricultural production, for example fruit tree plantations. It also excludes trees planted in agroforestry systems.
>
> *(www.fao.org/3/Y1997E/y1997e1m.htm#bm58}, accessed 14 June 2019)*

The Center for International Forestry Research in Bogor initially found the concept of forest so slippery that its leaders agreed not to define it. My silviculturist colleague, Ravi Prabhu, suggested (and we adopted) the following usage for our work on criteria and indicators for sustainable forest management in the 1990s: We differentiated a 'forest rich' environment, which was 'islands of people in a sea of forest' from a 'forest poor' environment, 'islands of forest in a sea of people' (Colfer and Byron 2001, p. 28). I continue this usage in this book. Unlike FAO, I explicitly *do* include here forest fallows that are part of swidden agroforestry systems (e.g., in Chapters 4 and 5) and do *not* include oil palm plantations (especially Chapter 7).

The idea of a landscape mosaic is another useful concept. It recognizes that forests are part of broader landscapes that may be composed of varying land uses, such as old growth, secondary and degraded forests, grassy areas, villages (and villagers), agriculture, roads and infrastructure (e.g., Mertz et al. 2012; Sayer et al. 2013; and the collection by Colfer and Pfund 2011).

The masculinities examined in this book were associated with areas where old growth dominated when the research began (1975 on Washington State's Olympic Peninsula, 1979 in East Kalimantan and 1983 in West Sumatra). Recently (2017) on the Olympic Peninsula, replanted forests predominate, with some old growth; and in 2019, in East Kalimantan, oil palm predominates, with secondary forest/agroforestry mosaics only near rivers' edges (see Figure 1.1).

My emphasis on reflexivity as essential in this book's 'journey' has meant the inclusion of some experiences and research in predominantly arid (Turkey, Oman), agricultural (the American midwest) and urban (Indonesia, US) areas, as well as occasional reference to masculinities in forests in Europe and tropical Africa, Latin America and other areas of Indonesia. I provide snapshots of typical forests, as they are encountered, in each setting discussed.

Framing of masculinities

As I analyzed my observations on masculinities in different contexts, I grew increasingly dissatisfied with the ways masculinities had been conceptualized in the literature (e.g., too focused on problems, biased toward atypical sexualities, overly influenced by western ideas about gender). I wanted to capture the

FIGURE 1.1 The research sites discussed in this book.

variety, the stability of cultural systems and the choices available to individual men in terms of their own identities. While considering these issues, I read Spiller's book *Erotic Triangles*. *Erotic Triangles* focuses very tightly on the role of dance in Sundanese[12] men's sense of their own masculinity. Although forests are not mentioned in his book, the Sundanese are sometimes referred to as 'mountain people', and much of their world is indeed forested. Spiller argued that masculinity among the Sundanese could be presented as a triangle, with one side of the frame the dancing men, the second side the *ronggeng* (a beautiful woman seen as part whore, part goddess) and the third side, the drummer who sometimes leads, sometimes follows the dancing men's creative yet constrained lead.

The harp: its structure, context and change

Here I ask the readers' forbearance, as I put forth my unusual approach. I have adopted Spiller's idea of a triangle, a constraining boundary, representing a cultural complex of ideas and practices pertaining to masculinities. I began to think of this triangle as being like a harp, from which men[13] could choose first, sets of strings (qualities, interests, norms) that form chords (bundles or clusters of strings, linked qualities/preoccupations), and eventually the 'song' that represents a given man's version of masculinity within his own life—represented on the cover of this book. The harp itself represents the stability and constraints that cultures provide, and the strings represent the elements of choice for individuals, with some chords dominant (more highly valued) within a given culture.

In modern times, a harp is an unusual analogy for a study of masculinities, given its feminine connotations in the West/North.[14] However, my hope is that this concept, slightly jarring to westerners, can help to remind the reader that we are looking at masculinity through a [this] woman's eyes—one important way to see masculinities in our hunt for understanding of gender dynamics.

Deciding to use this analogy derives from a problem all gender scholars have: How to acknowledge the stability that a bio-socio-cultural system maintains, its coherence and resistance to changes in many spheres of life, with the fact that nonetheless cultures do change, sometimes very quickly, and that individuals have equally obvious agency. The 'normative climate' proposed by Petesch et al. (2018) captures some of the stability of the harp, but here I'm seeking a clearer mechanism by which to reflect agency. Some normative climates are more open to experimentation and thus more conducive to an agency that not *only* struggles against constraints (as characterize many of the examples in Petesch et al.);[15] cultural harps also provide opportunities for agency, to varying degrees for both men and women.

Men, like women, are constrained by the choices and structures available locally, just as a harp constrains what music one can produce in a very general way. A Kenyah Dayak did not, until quite recently, have the option to be an academic scholar; an American logger would not likely have the option to take up Sundanese dance. But within the repertoire available, there can be considerable freedom to fashion one's identity and one's life. The amount of such freedom also differs from one culture to another. Connell (1995) and Pascoe (2007) for instance, show the formidable pressures American cultures can bring to bear in enforcing heteronormativity.[16] Paulson (2016) shows the system-based oppression inherent to interactions between some Latin American masculinities. Sutlive's (1991) collection on Borneo presents contexts far more conducive to men's experimentation with varying identities than do these examples.

One's choices have different implications for prestige within different groups of like-minded folks; the American village discussed in Chapter 3, for instance, included two groups or subcultures in the 1970s ('Locals' and 'Public Employees', Colfer and Colfer 1978). The son of a logger (a 'Local') on Washington State's Olympic Peninsula would have reaped disdain from his parents and their friends if he had opted to become a school teacher; this profession was seen locally as unmanly and marked by low salaries. In the same community, the son of a school teacher, a 'Public Employee', would have been—indeed, still would be—discouraged from becoming a logger, a profession seen by Public Employees as insecure and low status.

Different career choices there tended to imply different clusters of valued qualities. A logger, for instance, could pluck one set of strings symbolizing physical strength, practical skills, a love of heavy equipment and courage, and another set of strings related to independence and control in his home life, thereby creating his own song of manhood. A school teacher's song could be created from strings symbolizing academic knowledge, punctuality and security, on the one hand, and companionate marriage and being a responsible breadwinner on the other. But these are just 'ideal types'; any individual man can vary his choice of strings, with fairly predictable implications for his prestige within this or that group. A logger without physical strength, courage or practical skills could expect to be disdained; whereas a school teacher without those qualities would suffer far less. Conversely, a teacher whose academic knowledge was perceived to be inadequate may not have been tolerated, whereas a logger with this 'deficiency' would have been comparatively unaffected. Sometimes strings are plucked with forethought, a lifetime of commitment to that string, and/or full awareness of its implications for status and identity. Other times, people may act 'on the fly', without such forethought and with unpredictable effects. Still other times, one can imagine that a man might make conscious decisions to alter his identity and pluck a particular string accordingly, come what may.[17]

Since different groups assign different values related to the perceived masculinity of different qualities, one might cluster those qualities most closely aligned with masculinity within a given group, as the deeper tones of the harp, perhaps where the strings are longer and made of heavy wire rather than catgut. I focus in this book on the selection of harp strings, leaving the more complex chords and songs for future analyses.

In contexts where femininity is seen as the *opposite* of masculinity, qualities perceived to be feminine could be placed at the higher tones. This is, however, not a universal opposition (see Chapter 4); gender polarity is only perceived among some groups. Paulson (2016) agrees. Her thoughts on the power behind northern views are germane:

> Lugones [2007] urges readers to recognize that biological dimorphism, woman/man dichotomy, heterosexualism and patriarchy are not universal features of human social life but historically specific manifestations of one model—that which she calls the 'modern/colonial gender system'.
>
> The hegemonic power of this model is rooted in the interdependent development of colonial/modern gender systems and colonial/modern sciences that built a powerful association between a particular binary organization of gender and biological understanding of the sexual dimorphism of Homo sapiens. In the realm of gender, like that of race, the establishment of a knowledge/ideology that construes the dominant cultural order as biologically determined has proven extremely effective in leading actors to perceive that particular cultural order as inevitable and universal.
>
> *(p. 143)*

This book strives to incorporate greater attention to change,[18] as well as stability. Over the course of my lifetime (and in the material in this book) an incredible expansion of communication has occurred. This has expanded awareness, in the most remote parts of the forested world, of different attitudes, behaviours and ideals about masculinity, different 'harps', with a resultant expansion of the strings available locally. Such awareness requires greater attention to the issue of scale.

As early as 1960, Redfield (1960) had identified the relations between what he called the great tradition and the little tradition in peasant societies.[19]

> The great tradition is cultivated in schools or temples; the little tradition works itself out and keeps itself going in the lives of the unlettered in their village communities.
>
> *(pp. 41–42)*

Although many forest communities would not have been considered 'peasants' in anthropological usage at that time,[20] interactions between small forest communities and such outside influences were already present and have intensified enormously. Rather than one 'great tradition' impinging on people's lives as Redfield observed in Mexico, there are now many, via social media, email, television and radio, reflected in a May 2017 interview with an American 'Local' man from the forest community discussed in Chapters 3 and 7, see p. 324). He explains how social media have widened people's perspectives and acceptance, for instance, about homosexuality.

If we follow the harp analogy, with each community represented by one or more particular 'harp', we can see that people's awareness of harps from other contexts has grown exponentially. In Long Segar, East Kalimantan (Chapter 4), even in 1979, the people were exposed directly to masculinity-related influences from the church, the mosque, the government and international private industry—all operating on broader scales, with new ideas and more power, prestige and resources than local people had.[21] These influences had grown more intense 40 years later, with improved communications and infrastructure. We can consider the harps available—external and local—to constitute an 'orchestra' of harps, which carry the potential to produce beautiful music . . . or cacophony.

Other scholars have provided insights of relevance to this book:[22] Raewyn Connell, one of the earliest and most widely cited of masculinity scholars and the originator of the focus on hegemonic masculinity, has the unique advantage of having looked at masculinities from both a man's and a woman's perspective, having had a sex change late in life. In one of her earlier works (1995), she posits four kinds of masculinity, summarized conveniently by Pascoe (2007):

> R. W. Connell argues that men enact and embody different configurations of masculinity depending on their positions within a social hierarchy of power. *Hegemonic masculinity*, the type of gender practice that, in a given space and time, supports gender inequality, is at the top of this hierarchy. *Complicit masculinity* describes men who benefit from hegemonic masculinity but do not enact it; *subordinated masculinity* describes men who are oppressed by definitions of hegemonic masculinity, primarily gay men; and *marginalized masculinity* describes men who may be positioned powerfully in terms of gender but not in terms of class or race.
>
> *(p. 7, italics in original)*

These differentiations are recognizable in some places and times, but not particularly in others. Much of Connell's work is both insightful and fascinating but depends on a social context that strongly differentiates men and women, which is not the case, for instance, in some areas of Indonesia (see e.g., Chapter 4 or 7, this volume; or Cornwall 2016, more generally). In Chapter 3, we see two versions of what might be termed hegemonic masculinity, operating side by side in the same village. They reflect the ideological struggle over hegemony[23] discussed by Gramsci, especially in his prison notebooks (Forgacs 2000).

Gibson presents another way of looking at these choices, most clearly in Gibson (n.d.). He documents three different ways, linked to gender, that one can emphasize one's values, motivations and appropriate behaviour among the Makassar of Ara, South Sulawesi. These orientations are linked to the house (built on indigenous social structure), the mosque (linked to Islamic mysticism) or the school (emphasizing the Indonesian nation-state). In Gibson's later work he expands on these as modes of sociality with differing ontologies.

Lentz (2017), building on von Benda-Beckmann's (2009) notion of 'symbolic universes', discusses what appears to be the same three components—termed Traditionalism, Islamicism and Modernism—operative in Sundanese lives (p. 46).

In my terms, men choosing each of these modes of sociality (or identities or universes) would reflect different chords, or clusters of harp strings. In southern Sulawesi, Graham-Davies (2004) describes local recognition and spiritual relevance of five divergent genders, of particular interest as an intriguing example of the global divergence in forest-based masculinities that I hope to show in this book (see also Gibson 2005).

Gibson (n.d.) notes that "every individual in the world participates in several institutions that operate according to incommensurable forms of temporality, spatiality, subjectivity and authority" (p. 26). He shows the various ways in which local practice brings these divergent ontologies or modes of sociality together (e.g., a religious sheik [Islamic mosque] buried with an aristocratic wife [Austronesian house]).[24] I take from this observation (though he does not say this) that individuals have differing options or interpretations available to them for potential use in different circumstances.[25] Where I have referred to 'songs' above, Gibson refers to the internal logic of a given identity as 'grammatical' (or presumably not).

Keeler (2017), another theorist on masculinity, posits a spectrum with autonomy at one end and attachment at the other, which he sees as capturing one of the most fundamental issues that men must resolve in their lives and choices and in their interaction with [the less valued] women. For the Burman monks he worked with, "autonomy constitutes an unrealizable but incontestable goal to which all males should strive[;] it enjoys hegemonic authority" (p. 229). He further links these concepts with sexuality and hierarchy.

> I have suggested that autonomy stands at one end of a spectrum that runs between it and attachment, and that this links it to the end of another spectrum that runs between superordination and subordination. In other words, engaging in sex without becoming personally entangled with someone looks like the behavior, or privilege, of the person who enjoys what might be termed, aptly if perhaps coyly, the upper hand: the agent endowed with greater power and superior standing, with greater freedom to maneuver; and so, autonomy.
>
> *(p. 222)*
>
> So to the extent that autonomy constitutes the solution to the dilemma of attachment, and autonomy enjoys a privileged place in hierarchical views of the world, giving up sex has to arouse maximal respect.
>
> *(p. 225)*

So different from the views I've encountered and discuss in the pages to follow!

Ford and Lyons (2012), whose collection focuses primarily on urban settings, recognize the varying kinds of masculinity, that these are subject to change and that they are not essential features of men's biology (that women too can manifest features often described as masculine). These authors emphasize the performative rather than essential nature of gender:

> the construction of 'gender' as multiple and variable. Not only does such an approach provide a means to acknowledge female masculinities and male femininities, but it also affords a means to examine how men negotiate masculinities in their daily lives.
>
> *(p. 9)*

This view is consistent with the emphasis here on choice, agency (à la Kabeer 1999) and variability within cultural systems that may provide enough stability for human comfort and predictability.

The ethnographer as interpreter

In recent decades, as noted earlier, anthropologists have increasingly recognized the significance of their own world views and personal situations in influencing their interpretations of what they see. As a woman considering masculinity, this strikes me as particularly likely to play an important role in what I see, how I analyze it and what I report. Bringing these considerations to the fore, and being so honest about my own sexual involvements (discussed shortly) goes against established practice,[26] despite postmodern ideas about positionality and its relevance. Altork (1995) and Killick (1995) both address this issue clearly in their analyses based on their respective experiences in the US and Korea. Here I discuss my own peccadillos as well as my interpretation of their relevance for my findings and the ethical implications.

In considering the book, *The Wild Man*, by Schneebaum (2003)—which reportedly discusses his participation in homosexuality and cannibalism in the field—Altork says,

> One can't help but wonder if his writings would have caused more of an uproar if they had not been homosexually contextualized, i.e. if he had had passionate affairs with women 'natives' and written about it as candidly as he wrote about his homo-erotic experiences. It's almost as though he could be written off—the marginal writing about the marginalized—since his behavior didn't involve a cross-gender interaction. This is a sad indictment of one of the ways in which our culture compartmentalizes and privileges certain kinds of experience as being more meaningful and, perhaps, more worthy of being entered into the academic discourse.
>
> *(p. 135)*

This perspective is echoed by Bolton (1995), a gay man, who points out that gay researchers, already marginalized, lose less by discussing taboo subjects. He also argues that "refusing to share in sexuality across cultural boundaries helps to perpetuate the false dichotomy between 'us' and 'the natives'" (p. 140).

When Killick was asked to contribute to Kulick and Willson's (1995) book *Taboo: Sex, Identity and Erotic Subjectivity in Anthropological Fieldwork*, he was "bemused by the prospect of being seen as a 'token' straight white guy" and "puzzled by the task of writing about [his] love life in a way that would be honest and explicit, but would not read as either confessional, pornography, or an extended singles ad" (p. 79). I have struggled similarly.

Feldman and Shaw (2018), writing about the ethics of interpretive ethnography, discuss how "data are intimately tied to the perceptions, experiences, and meaning-making processes of the researcher". They highlight that such an approach

> ties data to the immersion of researchers in the everyday lives of those they study and is premised on integration of processes of data production and analysis. In this constitutive framing, the ethnographer is not separate from but, rather, is integrally linked to, and a part of, the story she/he tells.
>
> *(p. 6)*

I have no doubt that my own situations and the choices I've made about sexuality in my various field experiences have had an influence on what I've been able to understand about each version of masculinity. I therefore explain them briefly but honestly here.

My parents went to great effort to avoid inculcating in me any ideas of shame or disapproval of sexuality per se, though they warned me about the dangers of unplanned and premature pregnancy. As several recent feminist authors have argued (e.g., Fine 2010; Saini 2017), scientific interpretations—dominated by men's perspectives—about women's inherent passivity and choosiness in sexual matters have been greatly exaggerated. My own sexuality, particularly during my 20s and 30s when the sex drive tends to be particularly powerful, was definitely a force to be reckoned with. Logic and rationality did not always function perfectly when confronted with desire (and in the fieldwork context, curiosity).

In Bushler Bay (Chapter 3), I was in my 20s, sexually switched on, and married with one daughter (aged 3–6). My anthropologist husband and I were both interested in gender roles and in experimenting with both them and our relationship. We married in the 1960s, and by the 1970s, many were similarly experimenting, including with what was then called 'open marriage'. After a couple of years in Bushler Bay—a context where sexuality was ever-present and explosive—we both had sexual relations with another couple (secret from the community but discussed openly among ourselves). We attempted to live communally for a brief time, during which I insisted there be no more cross-couple sexual relations; my [conscious?] reason was that I wanted to be able to deny their existence honestly if community members should ask me about this issue. Jealousy (perhaps predictably) also raised its ugly head. Although there were personal implications for our respective relationships, I doubt that anyone in the community was injured by our experimentation. And I believe that our sexual involvement was helpful in understanding sexuality—a vital part of many men's masculinity—in that community.

By 1979, that marriage was coming to an end, and I went to Long Segar, East Kalimantan alone. I stayed for 11 months. After about five months, I became secretly sexually involved with my married field assistant, eventually deciding to have a child by him.[27] This liaison initially involved secrecy, dishonesty, guilt about his wife and several years later being tried and fined locally. Again, although I have regretted the impacts of this relationship on his wife (who eventually forgave me),[28] all in all the experience strengthened my understanding of local masculinity immeasurably. My then-husband came to Indonesia with my 11-year-old daughter during this affair, which I put on hold during his two-month stay. He knew of and was displeased about my relationship, though he had also become involved with someone else in the US. By the time my daughter and I returned home, he and I had agreed to divorce. I moved to Hawaii in 1980, and my paramour came with me for another year, further enhancing my experience of Uma' Jalan masculinity and other community features. Bolton (1995) has written about the 'false dichotomy' the sexual taboo creates between researchers and 'the natives' (p. 140), one that my own experience certainly minimized.

By 1983, when I moved to Sitiung, West Sumatra with my three-year-old son, I had become involved with an American fisheries biologist. He was starting a project on Java, and after a long-distance courtship (Java to Sumatra), we married in 1985. Shortly after that, we were joined by my teenage daughter. My new husband was monogamous by nature and I consciously decided to accept that practice as well. I was also part of a team, initially of three American men (later five) with their families and 10 to 15 Indonesian team members, almost all young, single men from Java. By this time, I was nearing 40, my own sexuality had moderated somewhat, and I was surrounded by religious, agricultural (and fisheries) scientists not—to my knowledge—inclined to social or sexual experimentation. My social context was very different, I was not lonely, nor was I tempted to 'stray'. The cultural system I encountered in Sitiung was also less appealing (and therefore probably less tempting) to me than was that of the egalitarian Kenyah.[29]

The differences in perspective related to these differing life conditions mean that I obtained different kinds of information in the different field sites. The topics of study also differed:

The Bushler Bay study, located in the Olympic National Forest and adjacent to the Olympic National Park, focused on the educational system, as part of a national programme (funded by the National Institute of Education), with one field researcher (all men, except me) in each of ten rural sites across the nation and a cross-site team in Cambridge, Massachusetts studying all ten, also all men. My first husband and I were job-sharing and monitoring what was envisioned initially as experiments in locally planned educational change, though federal involvement reduced local decision-making rather quickly. We were also writing our doctoral dissertations (discussed in Chapter 2, see also e.g., Corwin 1977; Herriott and Gross 1979).

In Long Segar, East Kalimantan, my project was funded by the US Forest Service, the Man and Biosphere Program of the UN and the Indonesian National Science Foundation (*Lembaga Ilmu Pengetahuan*). The project involved understanding the 'interactions between people and forests'. We were interested in the practice and ecological effects of swidden agriculture and the Indonesian resettlement programme, and the rationality of local people (which in those days—1979–1980—was still in question; see Chapter 4). Although I was part of a small, gender-balanced field team, we were spread around Kalimantan and I rarely saw the other team members. My occasional professional interactions were with men from the German Transmigration Area Development project along the rivers, in Long Segar and in the provincial capital, Samarinda, and with professors at Mulawarman University (all men introduced briefly in Chapter 6).

In Sitiung, West Sumatra, our project was funded by USAID and what I call the 'Soils Center'[30] on Java, managed by 'Island University' in Hawaii and 'Southern State University' in the American South. The project was initially designed to develop acceptable agricultural systems for the transmigrants[31] who had moved to West Sumatra from Java. We were implementing a participatory approach called 'Farming Systems Research and Development'. Initially we worked only with Javanese and Sundanese transmigrants; in the third year, we began studying and working with the long-resident Minangkabau as well. Our team grew from three to five, with the other researchers all married men. However, most of the spouses had professional capabilities (nutrition, nursing, social work, fisheries biology) and we regularly made use of their skills informally (Chapter 5).

Still married to my second (and current) husband, I began consulting for Forest Research Institute (FRI) in 1994. We moved to its headquarters, where I began as a principal researcher in 1996 and stayed until 2009—I remain a senior associate now. At FRI, my colleagues represented many countries, though the professionals were dominated by researchers from the global North (initially UK, France, Germany, Australia and the US) and support staff from Indonesia (mainly Javanese, Sundanese and Chinese Indonesians). My research focused on issues of people and forests: social aspects of criteria and indicators for sustainable forest management, adaptive collaborative management with forest communities, health and forests, decentralization/devolution of forest management and landscape management. Gender was important throughout. Partnership was ubiquitous, and in such contexts, I also worked with researchers from Africa, Latin America and other parts of Asia. Most were men. I had no more extramarital sexual liaisons.

Flow of this book

This book is organized chronologically, with the intention of showing the variation from place to place, but also taking the reader on a journey through changes occurring over the last 70 years or so. In Chapter 2, I introduce my own childhood, youth and young adulthood (circa 1950–1972). As mentioned previously, this allows the reader to assess for him/herself the likely biases and lenses through which I view masculinity—setting the stage for the chapters to follow, but also providing glimpses of the regional variety in American masculinities. Biophysical scientists will not be accustomed to such disclosure or its relevance. They may choose to skip this chapter. However, the variety of masculinities is further exemplified there. The style in which each subsequent chapter is written differs, reflecting the differing circumstances of the research, the different methods used and perhaps the differing times—but all addressing masculinities in forests.

How I've dealt with anonymity differs from one chapter to the next. In Chapters 3, 6 and parts of 7, I take the need for anonymity seriously. In the other chapters, because of the attitudes of the people involved, I do not. Chapter 2 is quite historical so only occasionally is anonymity important, something I try to honour as needed. In Chapters 4 and 5, people take pride in their cultural systems and would be disappointed to be disguised in a book such as this. In Chapters 1 and 8, there is little in the way of personal information about individuals who might be recognized. In some chapters, where anonymity is important, I disguise some dates by merely indicating the decade. In addition to conventional citations, I include material from my fieldnotes (called 'notes') and from my personal diaries (called 'journals') in the text.

Chapter 3 takes us to a rural American community in the Pacific Northwest in the mid-1970s. Ethnically, the community was almost 100% white and considered 'under-privileged' by outsiders.[32] About half of the men were engaged in natural-resource-based private industry (especially logging, the group A. M Colfer and I called 'Locals') and about half were employed in public institutions (the US Forest Service, the US Park Service, the public school, etc.; 'Public Employees'). This employment-based social structural differentiation accounted for much of the difference in masculinities, with very different harp strings selected by each group. Age also made a difference. Gender differentiation was extreme among Locals, less so among Public Employees.[33]

In Chapter 4, we move to the Indonesia of 1979 and 1980, where we encounter first, and superficially, the Balinese. After two months there, I moved to Long Segar, to a group of Uma' Jalan Kenyah Dayaks (referred to henceforth as 'Kenyah'), whose attitudes and behaviour relating to gender differed even more dramatically from those in the rural American logging community. There were identifiable masculine harp strings, but their significance in daily life was much less marked than in the rural US. And the chords that Kenyah men tended to play drew on different harp strings than those in the US; gender, as social structural and ideational systems, was muted (Ardener 1975).

In Sitiung, West Sumatra (Chapter 5), there was a complex ethnic mix: three Indonesian ethnic groups (Minangkabau, Javanese and a few Sundanese) intersected with professionals from the US (all white men) and Indonesia (also Javanese, Sundanese and Minangkabau) in the mid-1980s. The Minangkabau were a matrilineal and strongly Islamic group who practised swidden agroforestry and paddy rice agriculture; the Javanese were bilateral, less overtly religious and focused on farming field crops (rice, corn, soybeans); and the Sundanese, also bilateral but showing more dominance by men and a stronger Islamic orientation, preferred vegetables, fruits and fish farming where feasible. Masculinities differed among these three groups as well as among the Americans.

In Chapter 6, the discussion shifts from formal research results to an analysis based on four decades of personal experience working mostly with an international coterie of what Connell (1995) calls 'Men of Reason' (pp. 164–182). Rather than drawing on studies per se, it relies on my diaries and personal memories of interactions that imply the plucking of particular harp strings and chords.

Chapter 7 is divided into three parts. The first provides a return to the reflexive approach in Chapter 2, bringing the discussion up to the present in rural and forested upstate New York. The second part reports the results of a brief restudy of Bushler Bay in 2017 and the third discusses change among the Kenyah and in their forests, deriving from another brief restudy in 2019.

The conclusions (Chapter 8) summarize important differences in men's links with forests. The differences among Bushler Bay's Locals and Public Employees, the Kenyah Dayaks of East Kalimantan, the Minang, Javanese and Sundanese of West Sumatra and the global forestry elites whose decisions affect such peoples are revisited. I emphasize the implications of these masculinities as they vary by time and place, for forests and forest management.

Notes

1 Putz and Holbrook (1988) introduce western notions of such habitats thusly:

> Say the word *jungle* and one conjures up a vision of riotous impenetrable vegetation, drenched with steam and mist, teeming with wondrous, unfamiliar, and perhaps dangerous beasts. Say the word *jungle* again and the view shifts to an idyllic setting of palm trees and jewel like flowers in which the gentle inhabitants live in harmony with their surroundings. A third time and now the tropical forests form the backdrop for scenes of adventure, conquest, and discovery.
> *(p. 37)*

2 This excerpt was provided to me by Maureen Reed, who read an earlier draft of this book.

3 Briefly, power relationships based on sex/sexuality; how relationships may be organized and perceived (see also Glossary).

4 Hegemonic masculinity refers to the practices that legitimize men's dominant position in society and justify subordination of women and of men with non-dominant qualities and practices. Initially in the masculinities literature it was discussed as a culturally idealized form of manhood that was socially and hierarchically exclusive, concerned with breadwinning, was anxiety-provoking and differentiated (internally, hierarchically), brutal and violent, pseudo-natural and tough, psychologically contradictory and thus crisis-prone, economically rich and socially sustained, as well as applying to all men (summarized from Donaldson 1993).

5 Noted also by Spall (2016) in Angola: "As in many other contexts in Africa . . ., it was not obvious to men in veterans' families which style of masculinity was hegemonic, or if any was" (p. 162).

6 As Cornwall (2011) also complains of this literature, "We cannot see the men around us, the men who are part of our everyday lives, our friends, our allies in our struggles" (p. 202).
7 I took seriously Dove and Kammen's (2015) injunction to address 'the mundane'.
8 Cornwall, Karioris, and Lindisfarne (2016) note that they had each come of age in different times, shaped by the politics and ideas of the moment. Our intellectual trajectories had taken us along similar paths, but in quite distinct historical times" (p. xiii). Here I recognize that these contextual changes likely also influence my interpretations of masculinities.
9 Here, I use 'identity' to refer to how a person thinks of himself, what elements of his being he values and what differentiates him from others. These also apply to women.
10 These are studies in which data are taken from communities with little or no feedback of information or direct benefit to them.
11 Here, the emphasis is on men's concerns, but of course women's concerns are equally important and must be considered. Forest managers have not traditionally been systematically cognizant of the concerns of either gender.
12 Sundanese are the ethnic group originally from West Java. They also appear briefly in Chapters 5 and 6.
13 The degree to which this analogy can be used to reflect women's lives remains unclear. Here I focus on men, though it will become clear that many of the 'strings' men pluck can also be plucked by women, but often with different meanings culturally.
14 Wikipedia reports a long history of harp usage (which implies men's involvement), with contemporary symbolic connections with, among others, Ireland, the Catholic church and Guinness beer! (https://en.wikipedia.org/wiki/Harp#As_a_symbol, accessed 26 June 2019).
15 E.g., whereas I found the Javanese in West Sumatra (Chapter 5) to be happy to experiment with and adopt agricultural practices proposed by our agricultural project, they were rigidly devoted to their own health-related beliefs and practices, unwilling to consider 'modern' alternatives.
16 "Heteronormativity refers, in sum, to the myriad ways in which heterosexuality is produced as a natural, unproblematic, taken-for-granted, ordinary phenomenon" (Kitzinger 2005).
17 This discussion has benefitted from Chris Gibson and Susan Paulson's insights.
18 As Connell's (2005) later work notes,

> It is desirable to eliminate any usage of hegemonic masculinity as a fixed, transhistorical model. This usage violates the historicity of gender and ignores the massive evidence of change in social definitions of masculinity.
>
> *(p. 838)*

19 This section and the next may be of less interest to biophysical scientists, referring to social science theory and to reflexivity (my personal situation in each field site).
20 In the late 1960s, anthropologists still talked about 'primitive man' and 'peasants'. Using that now-abandoned terminology, this book deals with the 'civilized' (in Bushler Bay and my professional colleagues), the peasantry (the Javanese and Minangkabau) and the 'primitive' (the Kenyah). Understandings and vocabularies have changed.
21 Connell (2005) recognizes three levels of relevance for masculinity: the local, the regional and the global (p. 848). We must also remember Tsing's (2005) analysis of policy implementation, which shows dramatically how broader scale perspectives are altered and adapted locally in Kalimantan (reflecting Gramsci's hegemonic struggles).
22 Apologies to readers less familiar with social science terminology. The next few pages may be more difficult to read.
23 In Connell's (2005) terms, "Hegemony did not mean violence, although it could be supported by force; it meant ascendancy achieved through culture, institutions, and persuasion" (p. 832).
24 A similar meshing of these three elements is discussed by Kahin (1999) regarding the Minangkabau (Chapter 5).

25 Gibson (2018) does, however, conclude later

> that far from it being possible to identify some atemporal, pan-human mode of sociality, no stable mode of sociality or system of values can be identified within the activities of even a single human being.
>
> (p. 1)

26 Bolton (1995), a gay man, discusses the taboo on ethnographers engaging in sex in the field (p. 140), and argues persuasively against such a taboo.

27 In considering whether or not to engage in sexual relations with community members, one must balance one's personal system of ethics with one's interest in the deeper knowledge of the particular field setting that intimate relations afford, the potential to minimize the self-'other' distinction, the traditional disapproval of the academic community, potential harm to others, a possible reduction in the 'objectivity' some still consider truly possible in studying culture and one's own views on and experience of sexuality.

28 I later asked her why she forgave me. She said that her husband would have been having affairs with someone else anyway, and his relationship with me had at least resulted in material benefit to the family. I'd contributed to various community projects, sent money when she was ill, helped pay for her daughter in law's education.

29 Some cultures appeal to me more than others. The Kenyah kindness to each other, lack of an ideology of female inferiority, positive spin on hard work and flexible sex roles all appealed to me. At the other extreme was my time in the Sultanate of Oman, where men and women were strictly segregated in almost all realms, women were seen as inferior and dangerous, the initial response to any request was always 'no' and there was a harshness about interpersonal relations that I found difficult.

30 I use pseudonyms as needed throughout this book.

31 Transmigrants are participants in a longstanding Indonesian (and previously Dutch) programme to move people from densely populated Java and Bali to the 'Outer Islands' of Indonesia.

32 Although there are Native American communities on the western side of the Olympic Peninsula, there are none in the area of Bushler Bay, nor did I encounter Native American masculinities on this particular journey.

33 These folks more closely resemble the population imagined/described in much masculinities literature, which includes an assumption of strong gender differentiation and masculine fear of appearing feminine (see Morris and Ratajczak 2019, for a nice summary of such analyses as applied to violence against women).

References

Altork, Kate. 1995. "Walking the Fire Line: The Erotic Dimension of the Fieldwork Experience." In *Taboo: Sex, Identity, and Erotic Subjectivity in Anthropological Fieldwork*, edited by Don Kulick and Margaret Willson, 107–139. London and New York: Routledge.

Ardener, E. 1975. "Belief and the Problem of Women." In *Perceiving Women*, edited by S. Ardener. London: Malaby Press.

Bolton, Ralph. 1995. "Tricks, Friends and Lovers." In *Taboo: Sex, Identity and Erotic Subjectivity in Anthropological Fieldwork*, edited by Don Kulick and Margaret Willson, 140–167. New York and London: Routledge.

Colfer, Carol J. Pierce. 2005. *The Complex Forest: Communities, Uncertainty, and Adaptive Collaborative Management*. Washington, DC: Resources for the Future/CIFOR.

Colfer, Carol J. Pierce. 2018. "Before and After the 'Timber Wars': Context, Change and Potential Collaboration on the Olympic Peninsula." *Humboldt Journal of Social Relations* 40:137–162.

Colfer, Carol J. Pierce, and Yvonne Byron. 2001. *People Managing Forests: The Links between Human Well-Being and Sustainability.* Washington, DC: Resources for the Future/ CIFOR.

Colfer, Carol J. Pierce, and with A. Michael Colfer. 1978. "Inside Bushler Bay: Lifeways in Counterpoint." *Rural Sociology* 42 (2):204–220.

Colfer, Carol J. Pierce, and Jean-Laurent Pfund, eds. 2011. *Collaborative Governance of Tropical Landscapes.* London: Earthscan, CIFOR.

Connell, Raewyn W. 1995. *Masculinity.* Berkeley, CA: University of California Press.

Connell, Raewyn W. 2005. "Hegemonic Masculinity: Rethinking the Concept." *Gender and Society* 19 (6):829–859. https://doi.org/10.1177/0891243205278639.

Cornwall, Andrea. 2011. "Women's Empowerment: What Do Men Have to Do with It?" In *Men and Development: Politicizing Masculinities*, edited by Andrea Cornwall, Jerker Edström, and Alan Greig, 196–204. London: Zed Books.

Cornwall, Andrea. 2016. "Introduction: Masculinities under Neoliberalism." In *Masculinities under Neoliberalism*, edited by Andrea Cornwall, Frank G. Karioris, and Nancy Lindisfarne, 1–28. London: Zed Books.

Cornwall, Andrea, Frank G. Karioris, and Nancy Lindisfarne, eds. 2016. *Masculinities under Neoliberalism.* London: Zed Books.

Corwin, Ronald G. 1977. *Patterns of Federal-Local Relationships in Education: A Case Study of the Rural Experimental Schools Program.* Washington, DC: National Institute of Education.

Donaldson, Mike. 1993. "What Is Hegemonic Masculinity?" *Theory and Society* 22 (5): 643–657.

Dove, Michael R., and Daniel M. Kammen. 2015. *Science, Society and the Environment: Applying Anthropology and Physics to Sustainability.* London: Routledge.

Elias, Marlène, Netsayi Mudege, Diana E. Lopez, Dina Najjar, Vongai Kandiwa, Joyce Luis, Jummai Yila, Amare Tegbaru, Gaya Ibrahim, Lone Badstue, Esther Njuguna-Mungai, and Abderahim Bentaibi. 2018. "Gendered Aspirations and Occupations Among Rural Youth, in Agriculture and Beyond: A Cross-regional Perspective." *Journal of Gender, Agriculture and Food Security* 3 (1):82 107.

Enria, Luisa. 2016. "'I Must Stand Like a Man': Masculinity in Crisis in Post-War Sierra Leone." In *Masculinities under Neoliberalism*, edited by Andrea Cornwall, Frank G. Karioris, and Nancy Lindisfarne, 136–150. London: Zed Books.

Feldman, Shelley, and Linda Shaw. 2018. "The Epistemological and Ethical Challenges of Archiving and Sharing Qualitative Data." *American Behavioral Scientist*:1–23. https://doi.org/10.1177/0002764218796084.

Fine, Cordelia. 2010. *Delusions of Gender: How Our Minds, Society and Neurosexism Create Difference.* New York: W. W. Norton & Co.

Ford, Michele, and Lenore Lyons. 2012. "Introduction." In *Men and Masculinities in Southeast Asia*, edited by Michele Ford and Lenore Lyons, 1–19. New York: Routledge.

Forgacs, David, ed. 2000. *The Antonio Gramsci Reader.* New York: New York University Press.

Gibson, Thomas. 2005. *And the Sun Pursued the Moon: Symbolic Knowledge and Traditional Authority Among the Makassar.* Honolulu, HI: University of Hawaii Press.

Gibson, Thomas. n.d. "House, Mosque and School: Androgynous Transformations in South Sulawesi, Indonesia." *unpublished*:29.

Graham-Davies, Sharyn. 2004. "Hunters, Wedding Mothers, and Androgynous Priests: Conceptualising Gender among Bugis in South Sulawesi, Indonesia." PhD dissertation, Anthropology and Asian Studies, University of Western Australia.

Herriott, Robert E., and Neal Gross, eds. 1979. *The Dynamics of Planned Educational Change: Case Studies and Analyses.* Berkeley, CA: McCutchan Publishing Corporation.

Kabeer, Naila. 1999. "Resources, Agency, Achievements: Reflections on the Measurement of Women's Empowerment." *Development and Change* 30 (3):435–464. https://doi.org/10.1111/1467-7660.00125.

Kahin, Audrey. 1999. *Rebellion to Integration: West Sumatra and the Indonesian Policy 1926–1998*. Amsterdam, The Netherlands: Amsterdam University Press.

Keeler, Ward. 2017. *The Traffic in Hierarchy: Masculinity and Its Others in Buddhist Burma*. Honolulu, HI: University of Hawai'i Press.

Killick, Andrew P. 1995. "The Penetrating Intellect: On Being White, Straight and Male in Korea." In *Taboo: Sex, Identity and Erotic Subjectivity in Anthropological Fieldwork*, edited by Don Kulick and Margaret Willson, 76–106. London and New York: Routledge.

Kitzinger, Celia. 2005. "Heteronormativity in Action: Reproducing the Heterosexual Nuclear Family in After-hours Medical Calls." *Social Problems* 52 (4):477–498.

Kulick, Don, and Margaret Willson, eds. 1995. *Taboo: Sex, Identity and Erotic Subjectivity in Anthropological Fieldwork*. London and New York: Routledge.

Larson, Anne M., Therese Dokken, Amy E. Duchelle, Stibniati Atmadja, Ida Ayu Pradnja Resosudarmo, Peter Cronkleton, Marina Cromberg, William Sunderlin, Abdon Awono, and Galia Selaya. 2016. "Gender Gaps in REDD+: Women's Participation is Not Enough." In *Gender and Forests: Climate Change, Tenure, Value Chains, and Emerging Issues*, edited by Carol J. Pierce Colfer, Bimbika Sijapati Basnett, and Marlene Elias, 68–88. London: Earthscan/Routledge/CIFOR.

Lentz, Linda. 2017. *The Compass of Life: Sundanese Lifecycle Rituals and the Status of Muslim Women in Indonesia*. Durham, NC: Durham Academic Press.

Little, Jo. 2006. "Gender and Sexuality in Rural Communities." In *The Handbook of Rural Studies*, edited by Paul Cloke, Terry Marsden, and Patrick Mooney. London: Sage Publications Ltd.

Lugones, Maria. 2007. "Heterosexualism and the Colonial/Modern Gender System." *Hypatia* 22 (1):186–209.

Mertz, Ole, Daniel Müller, Thomas Sikor, Cornelia Hett, Andreas Heinimann, Jean-Christophe Castella, Guillaume Lestrelin, Casey M. Ryan, David S. Reay, Dietrich Schmidt-Vogt, Finn Danielsen, Ida Theilade, Meine van Noordwijk, Louis V. Verchot, Neil D. Burgess, Nicholas J. Berry, Thu Thuy Pham, Peter Messerli, Jianchu Xu, Rasmus Fensholt, Patrick Hostert, Dirk Pflugmacher, Thilde Bech Bruun, Andreas de Neergaard, Klaus Dons, Sonya Dewi, Ervan Rutishauser, and Zhanli Sun. 2012. "The Forgotten D: Challenges of Addressing Forest Degradation in Complex Mosaic Landscapes Under REDD+." *Geografisk Tidsskrift-Danish Journal of Geography* 112 (1):63–76.

Morris, Edward W., and Kathleen Ratajczak. 2019. "Critical Masculinity Studies and Research on Violence Against Women: An Assessment of Past Scholarship and Future Directions." *Violence Against Women* 25 (16). https://doi.org/10.1177/1077801219875827.

Pascoe, C. J. 2007. *Dude, You're a Fag: Masculinity and Sexuality in High School*. Berkeley, CA: University of California Press.

Paulson, Susan. 2016. *Masculinities and Femininities in Latin America's Uneven Development*. New York and London: Routledge.

Petesch, Patti, Renee Bullock, Shelley Feldman, Lone Badstue, Anne Rietveld, Wenda Bauchspies, Adelbertus Kamanzi, Amare Tegbaru, and Jummai Yila. 2018. "Local Normative Climate Shaping Agency and Agricultural Livelihoods in Sub-Saharan Africa." *Journal of Gender, Agriculture and Food Security* 3 (1):108–130.

Pini, Barbara, Berit Brandth, and Jo Little, eds. 2015. *Feminisms and Ruralities*. Lanham, MD: Lexington Books.

Putz, Francis E., and N. Michele Holbrook. 1988. "Tropical Rainforest Images." In *People of the Tropical Rain Forest*, edited by Julie Sloane Denslow and Christine Padoch, 37–52. Berkeley, CA: University of California Press.

Redfield, Robert. 1960. *Peasant Society and Culture*. Chicago, IL: University of Chicago Press.

Saini, Angela. 2017. *Inferior: How Science Got Women Wrong—And the New Research That's Rewriting the Story*. Boston, MA: Beacon Press.

Sayer, Jeffrey, Terry Sunderland, Jaboury Ghazoul, Jean-Laurent Pfund, Douglas Sheil, Erik Meijaard, Michelle Venter, Agni Klintuni Boedhihartono, Michael Day, Claude Garcia, Cora van Oosten, and Louise E. Buck. 2013. "Ten Principles for a Landscape Approach to Reconciling Agriculture, Conservation, and Other Competing Land Uses." *PNAS* 110 (21):8349–8356.

Schneebaum, Tobias. 2003. *Wild Man*. Madison, WI: University of Wisconsin Press.

Spall, John. 2016. "Fatherhood and Intergenerational Struggles in the Construction of Masculinities in Huambo, Angola." In *Masculinities Under Neoliberalism*, edited by Andrea Cornwall, Frank G. Karioris, and Nancy Lindisfarne, 151–164. London: Zed Books.

Sutlive, Vinson H., ed. 1991. *Female and Male in Borneo: Contributions and Challenges to Gender Studies*. Williamsburg, VA: The Borneo Research Council, Inc.

Tsing, Anna Lowenhaupt. 2005. *Friction: An Ethnography of Global Connection*. Princeton, NJ: Princeton University Press.

von Benda-Beckmann, Keebet. 2009. "Balancing Islam, Adat and the State: Comparing Islamic and Civil Courts in Indonesia." In *The Power of Law in the Transnational World: Anthropological Enquiries*, edited by Franz von Benda-Beckmann, Keebet von Benda-Beckmann, and Anne Griffiths, 216–235. New York: Berghahn Books.

2

[REMEMBERED] AMERICAN MASCULINITIES FROM MY GIRLHOOD (1940s–1970s)[1]

This chapter sets the stage for the subsequent ones by both clarifying my own positionality on the one hand and establishing a basis for comparison with forest masculinities on the other. I take the reader through my childhood and early adulthood, pulling out anecdotes and characteristics that reflect my experience with primarily American masculinities. Those features I learned as a child to associate with masculinity alerted me to what I saw in other cultures and in forests, both in terms of similarities and differences. Because of the interconnections between men's and women's roles, ideals and practices, I discuss both. I also provide a brief foray into Turkish and Persian masculinities, as my first exposures to more obviously differently gendered worlds.

These notions of masculinity touch on four likely types of harp strings, which emerge in all the cases to follow, but with differing valences:[2]

1 Ideology (man as legitimate controller of family, as superior)
2 Characteristics (strong, capable, outdoorsy, protective, courageous, patriotic, sexual, unemotional and potentially jealous and violent)
3 Interests (sports, vehicles, 'manly' academic disciplines)
4 Gender roles (employment, childcare, housework vs. breadwinner/provider)

Childhood

My childhood did not take place in forests. My father, my first introduction to masculinities, considered nature harsh and unforgiving, seeing it as a dog-eat-dog world, where the strong survive and the weak perish. His American frontier orientation saw nature as something to tame. My subsequent love affair with the forest emerged out of a childhood in very different habitats.

I was born in Melrose Park, Illinois, a suburb of the vast metropolis of Chicago on the shores of Lake Michigan ('Middle America') in August 1945. We then moved to the arid land of Norman, Oklahoma, regularly visiting the tiny village of Cyril (population ~1,000). Very few trees dotted this barren landscape, famous for its devastating dust storms during the early to mid-20th century. The next move, in 1950, was to Bloomington, Indiana (adjacent to Illinois). Indiana was wetter, but also an area with sparse tree cover. The landscape I recall most vividly was miles and miles of corn fields, as far as the eye could see—a 'forest poor' environment. I do, however, remember Brown State Park, which was forested. This current description probably reflects the 'original' forest type, of which there were remnants here and there:

> The most important tree in the park is the stain (*Cladrastis lutea*). This tree typically does not grow further north than central Kentucky, and has been designated as a state threatened species in Indiana. Other trees found in the park include at least four types of oak (black, chestnut, red, and white) and three types of hickory (bitternut, pignut, and shagbark). The park also contains at least two [sic] types of maple trees: black, sugar, red and silver. Patches of paw paw trees can be found throughout the park, and these trees produce an edible fruit. In areas with good moist soil, the black walnut tree grows, and this tree is an excellent source of wood for lumber or furniture. Among other trees growing in the park are the American beech, basswood, black cherry, black gum, and red elm.
>
> *(https://en.wikipedia.org/wiki/Brown_County_State_Park# Flora, accessed 30 May 2019)*

I enjoyed our periodic visits to the park, where we picnicked, and I sought crawdads in the creek.

But I was not a tomboy. What I really wanted to be at that time was a princess, to be found or rescued by a handsome prince. The gender relations portrayed in fairy tales were my standard fare: My knight in shining armor, the climber of ice mountains to win my hand, even the toad who would turn into a prince when I kissed him. Another handsome prince struggled through dense foliage that had magically grown up around a castle in which I, the bewitched princess, had lain sleeping for decades. Such images captured my imagination and formed a part of what I saw in childhood as masculine possibilities.

Another part of me pitied men. I learned early that men were expected to take the initiative in dating and sexuality, something I imagined to be terribly difficult, a responsibility I found frightening. And I realized they were expected to be economically responsible for families in ways women were not. I was grateful to be female.

I was interested in sexuality from a very young age, not having been shamed into hiding it. When we girls played a game in which one's future was told, I always hoped for 'love' (rather than 'fame' or 'fortune'), and the love I had in mind included being enfolded in the arms of a loving and manly man. Later such fantasies took on more overtly sexual tones.

Beside these delightful fantasies about what I considered then the upsides of masculinity and femininity, I also had the example of my parents. Until I was nine, both my parents were in college and then graduate school. Both worked, both took care of me, both studied. My father, Gene, made breakfast; my mother, Gwen, made dinner. But my father paid an emotional price for this gender equity: he often felt that he was not a proper man.

Dad was a product of the Depression (1930s), as manifest in arid, rural Oklahoma—with more oil wells and cattle ranches than forests. Born in 1924, he grew up in Cyril. Surrounded by a 'pioneer' culture where 'men were men', he felt he should be physically strong, capable, protective, honourable.[3] Eventually he adopted most of my mother's 'progressive' social and religious ideals, but he continued to remember—and sometimes be conflicted about—the Baptist teachings of his childhood (no drinking, no smoking, no dancing, no card playing, no sex before or outside of marriage). He never smoked, rarely danced or drank and suffered guilt about his sexual liaisons. Only card playing (not gambling) seemed to become truly acceptable to him.

Despite his supportive attitudes toward me and my potential future, he had some clear ideas about men's and women's natures. Once as a teenager in Ankara, Turkey (arid, urban), I remember asking him, on seeing the nude women in a *Playboy* magazine, if there were magazines with nude men in them. He expressed shock and said, "Oh no, a man would never do that". "Why," I queried? "He'd have more self-respect", was the answer.

Still, he completely supported both my mother's and my desire for education and interesting work. He shared my care when I was a child (born when they were both 21 and soon to continue college in Norman, Oklahoma). Later, after moving to Portland, Oregon, he mentored me regarding my profession (akin to his). After he retired, and my mother continued with paid work, he cared for his mother and then his father in law, when they grew old and came to live in my parent's household in Portland, Oregon (as 'forest rich' an environment as a city can be).[4]

He taught me the Boy Scout rules of behaviour—words I can still recite: "trustworthy, loyal, helpful, friendly, courteous, kind, obedient, cheerful, thrifty, brave, clean and reverent"—and he tried to live by them. He taught me what his Oklahoma grandfather had taught him: "Never do anything that you don't want printed on the front page of the newspaper". [I suppose rather than conform to society's expectations in this way, I opted to expand what I was willing to share about myself.] Dad learned when he was in his mid-30s that he was illegitimate—when his legal birth certificate came in the mail with 'illegitimate' stamped across it. He'd always thought his mother was divorced (also scandalous at that time in small-town Oklahoma).

Dad never, to my knowledge, explicitly told my younger brother, born in 1959, to be masculine, though Dad did show him how. He conveyed his antagonism to gay men (in contrast to gay women, whom he found amusing). Throughout my brother's childhood and youth in well-forested Portland, Dad encouraged his interest in sports in a way he never had with me (I wasn't interested), including watching it together on TV, attending my brother's games enthusiastically and praising him for his good shots (basketball)[5]—though he also praised me for anything I did that he found admirable (writing a good paper, looking nice, being kind, taking initiative). He showered us both with love.

I remember discussing crime with him as a child. Seeing the world in black and white, I argued that one should never kill another person or steal from him/her under any circumstances. My father disagreed, saying that if my mother or I were starving, he would readily steal; and if he or we were being attacked, he'd (reluctantly) kill another person. He had a strong protective ethos (a dominant string on his harp), which he felt all men should share.

I think only once did I see my father cry, in 1957, when news came that his grandfather (who'd been *in loco parentis* from Dad's birth) had died. One of my maternal uncles suffered throughout life because of his inability to control his tears. American men, unlike the Kenyah of Chapter 4, were not supposed to cry, pure and simple.[6]

Although Dad volunteered to serve in World War II, he never saw combat; he joined up again for the Korean War. But his abhorrence at killing another human being conflicted so seriously with his sense that a man should serve his country that he had a nervous breakdown. He was kind to all, human and non-human. When I was hurt in some way, I sought solace from him. He was the sympathetic one; my mother was likely to tell me to suck it up, tough it out, be brave.

Once, when I was a teenager, I saw my father threaten my mother with physical violence, though I don't believe he ever *acted* violently toward her. I was spanked occasionally as a child, normally for reasons I understood, by both my parents. But masculine violence in my experience never exceeded an occasional spanking.

Despite Dad's desire to be brave (and expectation that he should be—courage as an important cultural harp string), he feared many things. Among these were sleeping alone and being in the dark. He was afraid to sleep on the beach in Turkey (*camels might step on him*, 1950s), and later, in the 1980s, to drive along a muddy logging road in Sumatra (*we might have two blow-outs with only one spare tire*)[7] and to sleep on the porch of my village home there (*a tiger might eat him and my mother*).[8] Meanwhile, my mother strove to instill courage in me.

Father's fears did not interfere particularly with his thirst for adventure. He wanted to 'see the world' and managed to travel widely. He was not drawn to forests. Throughout his adventuring, he was always attentive to his own comfort (something my mother also aided in, keeping stores of sugar and instant coffee in her purse to bring out in 'emergencies').

Although Dad had love affairs in his middle years, his love for my mother was obvious, as was his emotional dependence upon her. Early in their marriage, their sexuality was blighted: Until she finally had a hysterectomy in her mid-40s, intercourse was painful for her, and he was not one to force the issue. In my youth, I knew he suffered from unfulfilled longings. Only later did I learn of his infidelities. His masculine 'song' included an active sexuality harp string.

He took seriously his prescribed responsibility to support the family (another widely shared, culturally valued 'string'), though he appreciated the fact that my mother was also always employed. His own mother had always worked, in rural Oklahoma and later in Chicago's suburbs, eventually settling in urban Oklahoma City—no forest in her world, until her final decade, which she spent in Portland. There, she often expressed perplexity about the need to protect endangered species; she had little interest in the beauty of forests. Four-leafed clover was more her style.

My parents married during World War II (1944), a day before Dad turned 21. After World War II ended, they finished college together in Norman, Oklahoma and much of grad school in Bloomington, Indiana—both forest poor regions. After seven years in Ankara, Turkey—arid and barren—Dad became a college professor, work he loved but considered unmanly,[9] in Portland. He suffered from the feeling that he should be doing something physical, rather than intellectual; his field was anthropological linguistics—not even one of the 'manly fields' like physics or mathematics (see Chapter 3 for similar views).[10]

Perhaps his love of football helped him cope with this 'failing'.[11] He loved watching sports, football being his real passion. He associated it with honourable masculinity, felt it taught boys to be strong, to work together for a common cause and to play fairly. He was repulsed when a coach or a team member was found cheating in some way. He would have been horrified and saddened by the ethical failings now so widely publicized within sports.

Homosexuality also repulsed him. He linked it with the French, whom he considered degenerate and effete, partly because of their acceptance of men having mistresses (interesting in light of his own infidelities). He preferred German culture, with its stereotypical emphasis on punctuality, precision and discipline, which he was able to separate in his mind from the hated Hitler and Nazism.

He had learned as a child that a 'real man' protected 'his' women.[12] My mother tells a story reflecting his attitudes: Once at a party, not long after they were married, another man approached her with sexual intent. She was handling it on her own, and managed, by moving away, to forestall any serious problems. My father observed this and did nothing—an action about which he expressed real shame. His rural Oklahoma upbringing told him he should have punched the guy for 'messing with his woman'. He also knew that in his new, more middle-class persona, such violence would not be admired or condoned.

His conflicts about his masculinity remained for decades. Even in his 60s, not long before he died, he felt the sting to his masculinity when my mother unwisely told him about a sexual liaison she'd had many years before. He'd seen his own far more ubiquitous extramarital affairs in a very different light from hers. Besides a sense of betrayal, hers tainted her 'purity', something he'd always believed she maintained—sexual purity [no sex outside marriage] was part of his narrative of femininity, the flip side for him of masculinity.

He 'bought' the cultural prescriptions about masculinity, and he admired what he considered my own femininity (my youthful concerns about my appearance, a quiet manner, a womanly walk). But both he and my mother assumed I'd go to college, marry and prepare for a profession (in no particular order). They both wanted me to be independent and able to support myself when/if needed. When I finished grad school, I was therefore unprepared for the degree to which jobs were still allocated by gender. I'd been treated as an equal and consistently told that women could do/be whatever they wanted. Such was in fact considerably less true in the 1970s than it is today.

My father was raised in his maternal grandparents' four-room home in Cyril, Oklahoma with his mother, V.[13] Dad and she shared a bed in the four-room house until he was ten. His grandfather, A., was well-respected in the town, and had been a store owner and later the mayor. A. and his wife had participated in the Oklahoma land rush of 1889, in which vast areas of the Oklahoma territory were opened up to white settlers. The pioneer narrative of the time emphasized making wild land productive and civilizing its inhabitants. An additional effect, obvious now and parallel to present-day trends in other parts of the world, was the take-over of Native American lands and decimation of the indigenous way of life (cf. Chapters 4, 5 and 7).

A.'s store went bankrupt during the Depression, when, the story goes, for too long he let people who couldn't pay buy food on credit. His daughter, V. worked in the telephone office or the bank in their tiny village throughout my father's childhood. The fact that my Dad remained ignorant of his illegitimate status until his 30s seems a testament to the community's high regard for his family.

I thought V. beautiful—she had been a 'flapper' in the 1930s, and remained attractive, even in middle age. I picture her slim figure, dressed in high heels and a bright red translucent blouse, her hair dyed black and arranged in a fashionable do, walking down the path, between her Dad's cactus garden and the single peach tree on which my swing was slung, into their house. After my father left home, she married in the 1950s. The first was a drunk who died of cirrhosis of the liver; the second a pitiful man who insisted on his masculine right to control yet lacked the intelligence and emotional stability to cope well with life. He was an orphan, had suffered polio as a child, been wounded in World War II and suffered shell shock. She was not a good judge of men. And it never seemed to occur to her to question her prescribed responsibility to do all the domestic work after a full day of paid labour—something she did throughout her working life. With me, she was extremely indulgent, in sharp contrast to the feminist women in my life: my mother and maternal grandmother, E.

V.'s mother, a bit of a recluse in her old age, was tiny, had a hump back and though sweet, was not particularly bright. I see her in the 1950s, feeding her 40-odd outdoor cats, wearing a bonnet (a fashion from the 19th century) and a long dress she'd made herself from a dark cotton print festooned with tiny flowers. I still have patchwork quilts she made with remnants from these dresses. She cooked meals, usually cornbread, beans boiled with a hambone, hamburger patties, sweet iced tea in heavy glass goblets; and in summer, corn on the cob or some other vegetable from their extensive garden. She canned produce, kept the house clean and did handiwork (quilted, tatted, crocheted, sewed). Her husband, A., in his 70s and 80s when I knew him, took care of the garden, but he was most often visible sitting on the porch with his dog, watching the world go by. Although his control of the household was subtle, my mother tells me his wishes were obeyed. Foods were cooked to his liking, even foods prohibited by his doctor. The two lived quite separate lives really, only eating together. She generally stayed in the back of the tiny house, he in the front, by then sleeping in separate rooms. She had her cooking and sewing; he, the garden and the news.

My maternal grandfather, R., whom I very much admired, was a 'progressive' Congregationalist minister in a Chicago suburb until the 1950s, when he and my grandmother moved to the small Illinois town of Abingdon. He was well read, intellectual and interested in literature and current events—not the bread and butter of the desired masculinities of his era.[14] But even in the depression, when there was little money available, he dressed spiffily (always wore a suit, tie and white shirt with cuff links). He was not physically strong or athletic; he was afraid of water and could not swim. Yet he made fun of one of my mother's boyfriends, whom he considered effeminate, by privately referring to him as 'Pantywaist' (mispronouncing his Hispano-Filipino name Pantilla). My mother remembers him as sympathetic though, in his role as pastor to a group of gay men.

In his dotage (the 1970s), after his third wife died (as had the first two), he came to live with my parents in Portland. Just before he came, his prostate became a problem. While operating on him, the doctor decided to do a vasectomy, not having asked his permission beforehand (something that would be subject to legal action now). On hearing of this, he became irate, considering it an insult to his manhood, that he might indeed still want to father children (!). He was frail by then, in his 90s, definitely unable to achieve many of his masculine ideals—so different from the rural Oklahoma version. But he still expected dinner to be served to him. My parents both humoured him. I never saw him work in the kitchen or indeed perform any other housekeeping function. Then, in my late 20s and deep into radical feminism, I had many debates with him. He would defend the status quo, telling me how he put women on a pedestal, and how much purer women were than men. It really came down, in his mind, to his frequent refrain: "Men are the pitchers, and women are the catchers".[15] I didn't buy it.

His first wife, my maternal grandmother, E., had been pulled out of school by her parents when she was 13, so had little in the way of formal education. Yet she too was well-read, intellectually curious and opinionated. She was strong, a feminist, and her husband's views of women were no doubt part of the conflict that sometimes raged between them. Another part was his philandering. His fashionable clothes were dictated, he maintained, by his profession; but there was never money, particularly during the Depression, to fulfill the needs of other members of the family. My mother explained her own acquiescence to many of my father's wishes as related to her desire to avoid the conflictual marriage she'd witnessed in her natal home.

When I envision E., I see her sternness. She was heavy and wore conservative dresses, usually ready for church. The life of a preacher's wife was full of duties—coordinating meals and treats at the church, teaching Sunday school, helping out in myriad ways in the community. She was loving, but strict. Her disapproval came quickly with any ethical failing. She too did (or organized) all the domestic work in the house, protecting her husband from interruptions and distractions from his preaching duties.

Besides my mother, E. and R. raised three boys. E. particularly admired the masculine physique, and all three boys were good athletes, proud of their physical strength. None was rigid about sex roles, but each conformed to masculine ideals: The eldest had an exciting job as a spy and travelled the world; the second flew airplanes in his spare time; and the youngest sailed. The eldest frequently cooked dinner (and helped raise four girls). The second and his wife worked together in their business (he was a consulting social worker; she, his secretary). The youngest became a practising psychologist, also helping to raise four daughters; his wife returned to her nursing profession when the girls grew up. In each marriage, the husband had the main job that supported the family; two wives were in charge of most domestic tasks. The third managed completely on her own during her husband's frequent travels but switched, I am told, into a helpless mode when he returned. He then took over much of the running of the household.

In 1955, my parents and I moved to Ankara, Turkey—as arid as rural Oklahoma—where I was to spend the next six plus years. My own father's sense that a man should protect his family was multiplied a thousand-fold in the culture we encountered in Ankara. There, ideally no one should even *see* a man's wife, though that was not a possible goal for most people. My father worked in an office with other expats; the secretary was an attractive young Turkish woman. One day, he planned a trip to another part of town and innocently took the young woman with him to translate. Her husband learned of this and came storming into the office the next day, threatening to kill my father. Only with great difficulty was the husband mollified, eventually leaving without violence. Looking at his wife was bad enough, but actually going someplace unchaperoned with her was totally unacceptable. Jealousy, which middle-class Americans typically try to downplay, was seen by many Turks, both men and women, as indicative of real love. And family honour was also involved.[16] A woman's association with a strange man could bring dishonour to the family (a patrilineal and patrilocal institution). Neither women nor other men were to be trusted, actually, when it came to relations between the sexes.

A woman's virginity at marriage was another litmus test for the honour of a family's men. In my early teen years, still living in Turkey, I read the book *Marjorie Morningstar*, in which the heroine, an American Jewish woman, reluctantly confesses to her fiancé, just before their marriage, that she is not a virgin; in the story, this discovery haunts him for the rest of his life. This narrative, which now would be far-fetched, strongly affected me for years, strengthening my concern to remain a virgin for a time. Eventually, after returning to the US in 1961, I realized that I would not really want a man who cared that much about something so fleeting.

Although the rights of Turkish men to control women were typically so interpreted from the Koran, reinforced culturally and practised in most households, rural women did a great deal of the agricultural work. One could routinely see the women out harvesting the grain, while the men sat and drank tea in the coffeehouses.[17] This did not liberate them from the women's prescribed domestic responsibilities of cooking, cleaning, childcare, etc. My father was horrified to see the women working so hard while the men relaxed, and I was surprised. There were clearly different gender norms at work here. Ideas about women's inferiority (intellectually and morally) were also commonplace, something I'd not encountered to any degree by age nine. The idea that women's bodies had to be so completely covered up[18] was another surprise, justified as protecting men from women's tempting sexuality, something women were helpless to avoid. That men should be strong, protective and in control of their families were somewhat familiar harp strings, but the ways they played out in Turkey were very different from what I'd seen in small town Indiana and Oklahoma.

When I was 16 (1961), we returned to the US, to Portland, where I experimented with many relationships with young men. I thought about what kind of man I wanted for a husband. By then, I knew no prince charming was going to 'rescue' me (nor did I particularly feel in need of rescue); no frog was going to appear for me to kiss. But my ideas of masculinity were surely affected by my surroundings. At 18, I fell in love with a man I wanted to marry but knew I couldn't. He came from a working-class family and was an apprentice machinist in his grandfather's small business, which he would inherit eventually . . . I knew he would be a good provider. He was gentle with me; I was sure he'd never mistreat me. He was tall and strong enough, if not handsome. He spent his spare time fixing up a car, another 'manly' pursuit. He was not intellectually inclined, though I very much respected his talent for creating and fixing things. Most fundamental in my decision not to marry him: He would have to stay in Portland to maintain his business, and I wanted more than anything to travel. I had a passion to contribute, to 'save the world' (Li's 2007 'will to improve').

I went off to Princeton University, another 'forest-poor' setting, at age 19, where I was one of nine women students—the university was experimenting with co-education in 1964–1965. I fell in love again, this time with a man, M., most American parents would want their daughter to marry. He was tall, good looking, intelligent and headed for grad school at Stanford University in political science (a field a step or two closer to the 'masculine' end of the American masculinity-femininity continuum than anthropology—though still further than forestry, for instance). His own parents were academics as well, his mother teaching English and his father, music, at a college in Oxford, Ohio (a habitat rather similar to Bloomington). When I think of his masculinity, the word that comes to mind is suave. He was an excellent dancer, knew a great deal about classical music, exuded urbanity. He wore Ivy League blazers, and shirts that his parents paid me to iron. He didn't iron them himself, and his parents didn't expect him to. I stayed up all night with him one night typing his thesis, as he put on the finishing touches. He was skilled in the role of romantic suitor. He bought a convertible one summer and when I left town on a bus, he drove along beside it for several miles, waving and blowing kisses. He was a good baseball player, though not on a college team; he coached girls' softball during the summer—demonstrating his sports acumen as well.

I was still thinking of marriage at that time and felt some of the anxiety that other young women felt, wondering if he would ask me to marry him (girls weren't to pop the question in those days). When he invited me, in 1965, to come live with him instead, I was shocked and offended. It was literally only months before a sea change occurred, at least in my circle, pertaining to living together rather than marrying. Soon after, unmarried co-residence became socially acceptable, but not at that time. I refused. Then a few months later, he did ask me to marry him. Much to my father's surprise (and a bit of chagrin), M. asked him for my hand (to which my father responded that it was entirely up to me). M. gave me a diamond that had been his grandmother's.

However, as the engagement wore on, I began to have serious misgivings. I had a recurrent vision of myself being enclosed in a white picket fence, living a life of domestic misery. I knew that my appearance was key to his attraction to me, that he wanted me partly as an ornament of which he could be proud (an indication of his success vis-à-vis other men). This made me suspicious—how would he feel when my beauty faded? Was his love genuine or mere infatuation? I also knew that he expected me to conform to social norms much more closely than had been my habit or preference. In such a relationship, I wondered how much I could pursue my own professional interests. His choice of secure funding for three years at Stanford over the opportunity of a funded year in Germany was evidence that we had different attitudes about adventure. He was sensibly concerned about his future; I longed to travel. In the end, I broke off the engagement.

Young adulthood

My next love (Michael) and I met while we were studying anthropology at Portland State College in 1966. He was from a tiny settlement in Maine, where his family, poor as church mice, had lived on the well forested banks of a large river. His mother had borne 12 children (11 living then); his father was a carpenter and had strong convictions in favour of labour unions, resulting in frequent periods of unemployment. He was a Catholic patriarch in full control of his family life; his word was law at home, and women were explicitly inferior. Sexuality was a prime feature in the view of masculinity he conveyed to his sons. His failures in providing for his family were perhaps compensated by ideological commitment to a man's right to rule the roost. The girls and his mother did all the housework, though without much in the way of modern technology, there was plenty outdoors for the boys to do as well. Ideas of women's inferiority were explicit, and Michael now remembers his youthful fears of seeming or being feminine.

He and I spent the summer of 1968 with his parents, living down the road from his two sisters, both of whose husbands had adjacent farms. In these two families, the men's jobs were to work the farm. Michael's sister, M., had set some ground rules before agreeing to marry R.: she would work in the house, not on the farm, and she would have her own kitchen and a separate part of the large old farmhouse, which they shared with his parents (whose mutual antipathy, incidentally, was so great they had not spoken to each other for years). The other sister, P., had made no such agreement beforehand and, besides taking care of the house, was active in outdoor farm work as well. Both R. and his brother in law, L., were 'strong, silent types', L. the more outgoing of the two.

Years before, the two men had had a disagreement in which R. accused L. of flirting with his wife, and the two had stopped speaking. Sometime later, L's wife P. called R. for help: L. had been attacked by his own bull and was trapped against a pole, hanging onto the bull's horns for dear life. R. and his brother came with a gun, and L. pleaded with them not to kill his prize bull. They were able to get control of the angry bull, probably saving L.'s life. On another occasion, in L.'s absence, R. saved L.'s barn from burning down. But the two still would not speak.

Ideals of masculinity in this area of rural Maine included stubbornness (also called 'persistence'), a touchy pride and an unwillingness to compromise. Related to that was the 'courage of their convictions'. Michael tells me that both these men fought in World War II, both decorated for bravery. They, and his third brother in law, "were the men who taught me, by default, what it meant to be a man" (email 6 Sept 2019).

Michael grew up in this context and joined the military right out of high school. Although an outgoing and convivial man when I knew him, he was short (5'8") and wiry. He had a 'baby face', and an unusual, slightly high pitched, gravelly voice. He told me stories of his days in the Air Force, when he was quick to anger and got into many fist fights (in Germany and the Black Sea area of Turkey, both 'forest-rich'). Some of these fights were related to attacks on his masculinity, though he later joked that he eventually realized he could get free drinks by initially playing along with the many homosexuals attracted to him. He has also commented more recently that the gay servicemen could talk about something besides 'pussy[19] and football'.

Our married life took place largely in forest-rich environments, whether urban (in Portland and Seattle) or rural (in Bushler Bay, see Chapter 3). As students of social life at a time when radical feminism was very much in the news, we discussed our own relationship at length. Early on, as we contemplated marriage, he told me that he had no interest in 'taking care' of me. He did not want to be responsible for another human being. My romantic childhood notions still held some sway and I felt disappointed that this widely held American harp string was not part of his masculine song, that it held no purchase for him (despite my commitment to working). I still thought it would be nice if my husband *wanted* to take care of me, even if it was unnecessary to do so.

When I became pregnant in 1968, he expressed his strong preference for a boy from the very start and continued with such comments throughout the pregnancy. He would laughingly affirm his preference for 'boats, beer and boys' [in children]. However, when our daughter emerged instead, he accepted her enthusiastically. He also happily accepted my idea of dressing her in greens and yellows rather than pinks (or blues); we agreed to try to raise her with as few preconceived gender roles as we could.[20]

Sexuality was probably the most central element of his notion of his own masculinity. He was proud of his abilities in that sphere and used to joke that he wanted to die being shot by a jealous husband *in flagrante* (though he was actually not prone at the time to illicit affairs). He expressed his sexuality in private and in public, dancing suggestively, flirting with all the women present, showing affection easily. By the time I knew him, he had abandoned any violent tendencies, reacting with humour if at all, when someone made an offensive remark or tried to bait him. Like my father, he'd accepted the middle-class avoidance of such violence.

Another of his loves, probably related to his notions of masculinity, was an interest in boats. He had a passion for sailboats and learned both to sail and eventually to make beautiful sailboats. He fulfilled my own need for adventure in a way that my earlier fiancé would not have been able to. Michael hoped to sail around the world, and that was part of our plan for many years, though we only made it to Hawaii in 1977 in the Red Witch, a 37' schooner, the hull of which we managed to buy in 1975.

Meanwhile, in graduate school, I dealt with the masculinities of my professors and doctoral committee members, all men, who shared many values with my father and uncles. Although my committee expected me to perform as well as men intellectually, masculinist ideas about gender, sexuality and power occasionally raised an ugly head. One committee member, well known for his extramarital affairs and multiple marriages, felt comfortable kissing me passionately whenever we met. As a member of my doctoral committee, he had what felt at the time to be life and death power over me. When I realized his kissing would continue unless I protested, I screwed up the courage to remind him of my happy marriage and fidelity. Thankfully that stopped him without adverse effects on my academic career. I knew of another committee member's attraction to me (he confessed in 2018 that he'd been in love with me), which thankfully he kept in check, while engaging in other extramarital affairs he told me about.

In 1971, when we had finished the coursework for our PhDs, Michael and I moved to arid and forest-poor San Bernardino, California, where Michael had a teaching job.[21] Although San Bernardino itself was arid, with few trees, 40 miles to the east of town in the foothills of several mountain ranges, lay the San Bernardino National Forest.[22]

This move and associated changes prompted a lot more discussion about gender roles, as I was stuck at home taking care of our daughter, and he was off doing the professional work that I too longed to do. At the time, I read an article (one which I cannot now find) suggesting that married couples develop a contract on the sharing of domestic work, one that recognized that both housework and jobs typically included some undesirable tasks. I remember the anxiety with which I watched him read the article, fearing that he would not agree—which I recognized was likely to be a major rupture in our relationship. It made so much sense to me. When he'd finished, he sat silent for a short time, thinking about it, and then turned to me with a smile and suggested that we try it. What a relief!

The suggestion in the article was that first the spouses develop a list of tasks that needed doing. Each was then to write down her/his preferences about performing these tasks, and then divide the work in such a way as to minimize each person's involvement in tasks they disliked, as they developed a practical and equitable division of labour. This approach was helpful in minimizing social pressures to comply with gender norms—by making the contract explicit, written, and agreed upon by both parties. I remember my own sense of liberation as we began living by this contract.

We began taking turns having primary responsibility for our two-year-old daughter, which gave me time to work on the (then unpaid) professional tasks I wanted to do. Contributing, as this plan did, to my own professional development was also good for him, as he wanted the option not to be the primary wage earner sometimes. His having such an option required that I too maintain my professional marketability. I was also better able to ignore social pressure to cook, for instance, when we had decided he was to cook that day; similarly, Michael was able to ignore people's meaningful looks (a few years later) about mowing the lawn, since we'd decided that was one of my tasks. We spent a lot of time trying to ensure that our relationship was an equitable one. We were equally concerned about the prejudices and biases against working women on the one hand and the intense pressures on men to support a wife and children on the other. Neither seemed fair to us.

With this perhaps utopian world view and set of practices, we set off for Iran on a Fulbright-Hayes grant in 1972,[23] where I encountered other sets of masculinities. I'd anticipated that my experience in Turkey would prepare me for Iran. However, there were significant differences. In Tehran, the Persian men struck me as comparative dandies, dressing beautifully, using hand cream and behaving in ways I was not used to considering masculine.

On one occasion I was invited to a party where the relevant bureaucrat suggested we trade my sexual favours for his granting my husband's visa to stay in the country (an offer I rejected, with considerable anxiety). Far more than in Turkey, I was subjected to regular sexual harassment (pinching, patting, grabbing and kissing) whenever I ventured out on the street—despite trying out various kinds of attire, including the all-encompassing *chador*. The 'reason', I was told (as in Turkey), was that women are too tempting for men; men don't have the will to resist.

In Persian people's homes, with my husband, men treated me respectfully, though they were shocked to learn that my husband had paid no bride price for me (the lack thereof, evidence of my having no value, also a reflection on *his* worth). The fact that American homes typically had no garden within their walls was another shock: Such a garden was valued as a place of domestic pleasure, where women (and perhaps men) were safe from other men's eyes and bodies.

The one outfit that seemed to serve as effective protection in the public sphere—in Shiraz rather than Tehran—was tribal clothing; Qashqa'i women were reputed to carry knives and be capable of defending themselves. In the Qashqa'i village of Dowlatabad, where we stayed in July 1972, the men worked in the poppy and wheat fields; the women churned butter and spun wool in preparation for making carpets during the winter months, besides cooking, cleaning, and taking care of children. Our host, the old village headman, who had two wives, took on the task of training us in Qashqa'i ways. He spent a lot of time correcting our locally inappropriate behaviour; we regularly heard the phrase, *pis dır* ('That's bad'), assessing something we'd done.[24] In this rural and domestic world, young men behaved toward me with curiosity, showing no disrespect. We were taken on as pseudo-family members.

Values portrayed

Here, I summarize the elements of American masculinities that made their way into my world view during my early years. They may be more widely shared, though such a conclusion will have to be more systematically examined. The relevance of these patterns for this book, though, is as a backdrop against which I examine masculinities in other worlds.

Ideology

I knew men who considered women both inferior (e.g., ex-father-in-law) and simply different by nature (father, maternal grandfather), but the ideology that permeated my childhood and young adulthood was one of equality. Although my father believed that men and women are different, he still believed that women could perform most jobs equally well and should have equal opportunities. My parents, my husbands (and many of my maternal relatives) had strong personal commitments and convictions about making the world a more equitable place. Exposure to inequitable perspectives did not consciously affect my own expectations about masculinity in my youth. I expected to find a man who would support my professional goals and work with me as a companion and equal. My marriages have been 'companionate'.[25]

Love of place was evident wherever I lived. The men in forests tended to love the beauty of the mountains and trees, including the latter's commercial and aesthetic value; men in cornfields waxed eloquent about how far one could see and the agricultural productivity of the plains; city dwelling men appreciated the services (education, arts, health care, etc.) available in cities and some from all habitats were drawn to urban employment opportunities and felt the lure of 'city lights'.

Pioneer values, emphasizing *using* natural resources, taming the wilderness and manifesting independence, strength and ingenuity, were identifiable in all the contexts of my youth.

Characteristics

I was exposed to the ideas that men should be strong, capable, protective, courageous and patriotic from an early age—though women were also enjoined to be as well. Sexuality as important to men's identities emerged in my consciousness as I began to deal with my own sexuality. Beginning at puberty my sexuality took over much of my attention and identity (as is obvious in personal journals from my teen years). Marriage to a man who tightly linked his sexuality with his manhood made its importance to some men even clearer. And as I aged, I learned of my father's and grandfather's peccadilloes. I heard about jealousy and violence as characterizing some men, but I learned that these were undesirable, to be avoided; and they have not played a very important role in my life.

Strength and courage have proven important for men in all the forest-rich settings I encountered early on, perhaps because of the need for such qualities in less technologically sophisticated, extractive forest activities. The emphasis on protection of women and on patriotism and religiosity seemed to thrive more obviously in the forest-poor contexts.

Interests

Masculine interest in sports as a manly concern was exceedingly common in my childhood and young adulthood, both within my family and without. The beginning of football season marked a significant change in domestic routines, and Christmas or Thanksgiving dinners were scheduled around the important games that in earlier days coincided with the holidays. During football season, Dad ignored my mother almost completely. My brother followed the country's basketball teams assiduously until his death. My former husband stands out as having no interest in organized sports, but sailing—a more individual sport—has been central to his identity since his 20s.

Love of cars and fascination with heavy equipment also characterized many of the men of my family and beyond. My father regularly bought new cars, which he saw as emblematic of his status in life as well as simply something he loved. My first husband made the case for buying a pickup truck, with which he was utterly delighted, before we could really afford the payments. Turning a 37' aluminum hull into a sailboat intended to sail around the world, was another indication of this largely masculine love (boatbuilding, something he continues to this day).

My own family, as an academic one by and large, emphasized intellectual pursuits. Although I was quite aware of the perceived gender implications of different disciplines (math and science for men; arts and literature for women), many in my family were involved in the social sciences—a brand of studies less gender-differentiated. All my uncles and my mother studied psychology or social work; my father was an anthropologist; and several of my cousins (I only have female first cousins) followed suit.

Involvement in team sports, effectively combining cooperation and competition while also emphasizing development of physical strength, may well have played a significant role in preparing men for jobs such as logging, as argued by the men of Bushler Bay (Chapter 3). Different combinations of these concerns were useful in the formal forest management bureaucracy (Colfer 1975).

Men in forests seemed particularly prone to the love of the heavy equipment used in logging in the US, where the smallest vehicle of general masculine interest was a pickup truck. Many young men spent endless hours during my late teens and early 20s fiddling with engines, repairing vehicles and lusting after ever-bigger, more powerful machines. I suspect the same interest in rural New York in 2020, based on the ubiquity of pickup trucks among men I see coming to local diners. Forest management continues to make regular use of related skills.

Gender roles

Although the differences between ideal and real behaviour with regard to gender roles are clear, there remained patterns of behaviour, not expressed verbally, which provided potent examples reinforcing conventional gender roles. My mother cooked what my father wanted to eat, as did my grandmothers for their husbands. Mom usually complied with Dad's preferences in life, and when she didn't, he complained. When his mother came to live with them, his mother's wishes were prioritized over my mother's. When he obtained an interesting job, the family moved, and he was able to finish his PhD; Mother could not. I picked up the sense that men's wishes were more important, that paying attention to them was necessary. Perhaps as a result, I strove to be like my father, more than my mother.

The problems for forest management of adherence to these traditional gender roles has become more and more obvious. There are of course equity concerns when only men's work is recognized in forests. And as the management of forests has broadened beyond commercial timber extraction—to include non-timber forest products, biodiversity, forest and human health, the livelihoods and well-being of forest inhabitants—managers have begun to recognize that the views, knowledge and experience of women also have relevance. The US Forest Service has made impressive strides in involving women in the forest bureaucracy, but attention to women's concerns in communities has lagged. Additionally, the ideas about appropriate gender roles reflected here, from the mid-20th century, did not adequately prepare men for present-day forest management technology, goals or practices. Fewer and fewer jobs require the physical strength once so central to successful logging.

I close this chapter with reflections on my impressions of these clusters of ideas about masculinities in the US.[26] They reflect my own experience. I cannot ascertain the degree to which these ideas are more widely shared. The passage of time is also of interest: To what degree do these generalizations represent an earlier era (the 1940s—1970s), rather than real differences today (also discussed in Chapter 8)?[27] These perceptions set the stage for the discussion of similar observations in other very different contexts, where I conducted ethnographic research. These ideologies, characteristics, interests and gender roles are built from the strings of the harps put forth in Chapter 1. Each man can select clusters, which together form 'chords' and 'songs' of masculinity—bringing together individual choice with the cultural constraints implied by the harp itself. The choices men make and the options available in any given harp have different implications for relations with other men and with women. There is often a power dimension with significant, sometime adverse, effects.

In sum

Some qualities were defined as masculine in all three social contexts (rural Oklahoma, rural Maine and midwestern intelligentsia) discussed in this book: providing, competitiveness, assertiveness, control of self (no crying or emotionality) and others, sexuality and physical strength.

In my own background I saw the masculine values of rural Oklahoma—now visible in President Trump's political base: strong division of labour between men and women, with men strong, protective, but autocratic breadwinners. Patriotism and religious fervour were often part of this complex; as was love of sports and related competitiveness, acceptance of violence as part of life (including in defense of 'one's women') and admiration of courage and hard physical labour, ideally outdoors.

Rural Maine's version differed somewhat. Emphasis on the gendered division of labour remained as did the focus on physical strength and hard labour for men; but a strong emphasis on being true to one's commitments and convictions was also evident—somewhat equivalent to the religious convictions of rural Oklahoma. Overt male-dominated sexuality popped up more clearly in the Maine version I witnessed than in Oklahoma, and male protectiveness of women was less evident.

Midwestern intellectual masculinity (Ohio, Illinois) included a less strict division of labour, though men were still seen as ideally the breadwinners. Intelligence, education, 'proper' attire and sophistication (including adventure) took on greater importance for this masculinity. Hard physical labour and physical strength could fall by the wayside, but the relevance of sports often remained.

The accumulation of academic knowledge and wide-ranging experience, and the exhibition of 'Culture' (with a capital C) were ways to accrue prestige. These qualities, interestingly, were at least theoretically also accessible to women and thus could open up potentialities for gender equity.

In none of these ideal complexes did men in those years typically admit to participation in household labour and many simply did not do it. Some who did 'drew the line' at changing a diaper, doing dishes or sweeping the floors.[28] These tasks were beneath their (public) dignity. Yet such involvement was not uncommon in the households I knew well, so I wonder how many men spoke one line and acted another.

A given man could select from these various qualities his own preferred repertoire—tempered by the varying gender rigidity of his context, his own capabilities and his preferences. He would have to keep in mind the particular regional 'harp' in which these chords were strung—and his selection would be valued or disvalued accordingly by himself and by those he knew and loved.

In Chapter 3, 4 and 5, we shift gears into the conventionally ethnographic. The masculinities I describe next are based on systematic fieldwork over multiple years, including specific studies and resulting data. The divergences described reflect cultural, geographical and temporal variation—as well as my own evolving views, interests and experience.

Notes

1 I disguise the identity of some of my relatives who might prefer anonymity. For those whom I concluded would be unlikely to care, I used their names (checking in some cases with them). Special thanks to Gwendolyn Marie Harris Pierce (my mother), A. Michael Colfer (my former husband) and Michael S. Cummings (my former fiance) for their comments on an earlier iteration of this chapter.

2 It's worth remembering that *not* fulfilling some of these can also be seen as a negation of one's masculinity. This is particularly problematic in contexts where gender is polarized (see e.g., Chapter 3).

3 Gibson's (2013, 2016) analyses of the cowboy icon—very prominent in rural Oklahoma—and clothing suggest a meaning for little boys somewhat comparable to my own daydreams about being a princess. His analyses explicitly recognize the ambiguities within the cowboy stereotype and the alternative harp strings that can be twanged to produce this variety of masculinity.

4 Portland has the largest forested urban park in the US, at 5,200 acres (www.forestparkconservancy.org/conservancy/initiative/; equivalent to 2104.44 ha). The state of Oregon is 48% forested (total land: 63,000,000 acres or 25.5 million ha), with about 60% of forest land owned by the federal government (www.oregon.gov/ODF/Documents/AboutODF/ForestryFactsFigures.pdf, both accessed 29 May 2019).

5 Veri and Liberti's (2019) analysis of men's cooking outside football games provides recent evidence of the centrality of sports in some American masculinities.

6 I think this taboo has moderated in recent years, however.

7 Dad's images of the forest were reminiscent of those quoted in Putz and Holbrook (1988), from Joseph Conrad's *Almayer's Folly* 1895:

> and in the midst [of the giant trees] the merciless creeper clung to the big trunks in cable-like coils, leaped from tree to tree, hung in thorny festoons from the

> lower boughs, and, sending slender tendrils on high to seek out the smallest branches, carried death to their victims in an exulting riot of silent destruction.
>
> *(p. 37)*

8 Wignall (2016) describes similar conflict between bravery and fear in The Gambia.

9 Grindal (2011), who studied masculinity in Tallahassee, Florida notes a view that could easily have come from my father's hometown or from Bushler Bay Locals (Chapter 3):

> The boys at Down Home Auto Repair also have their stereotypes and refer to university professors, as well as state bureaucrats, as snobbish and uppity, rude and demanding, pussies who are not willing to fight for their country, and pointy-head intellectuals who couldn't work eight to ten hours with their hands if they tried.
>
> *(p. 99)*

McGinley's (2004) discussion of masculinities at work in the US portrays a context more similar to the university where my father found himself in middle and old age.

> Masculinities reinforce stereotypes of the proper role and behavior of women and men at work. . . . Some of these practices include aggressiveness, competitiveness, informal networking, and regarding women as sexual objects, caregivers, or 'aggressive bitches'.
>
> *(p. 365)*

10 Though, as my mother reminded me, he had gotten a bachelor's degree in mathematics with physics as a minor and his first publication was in an astronomy journal.

11 See Anderson (2009) for a thorough but unsympathetic discussion of the role sports plays in masculinity in the West. He describes significant and increasingly equitable changes in college age sportsmen's attitudes toward masculinity in the 2000s in the US and UK (see Chapter 7, this volume).

12 This idea may have been strongly stressed by his mother, as I learned on her deathbed that he was the product of rape—though her lifelong devotion to my father was clear.

13 V. told me late in her life that she'd never loved anyone except my father and me (though she married two men, both of whom died before her). She appeared to love my daughter as well, but less my 'dark' son (who was actually the same colour as me, with Kenyah Dayak paternity). She was a quiet racist, unconcerned about the environment, believing that money made one happy. She followed the law, even when she didn't agree with it. She believed that she would see her relatives on the other side of 'the pearly gates', which she claimed to have seen in a near-death experience earlier in life. Despite the discrepancies in our views, my father and I loved her very much.

14 My plural usage here recognizes that different contexts recognize different masculinities as locally hegemonic.

15 My husband recently pointed out to me the centrality of the catcher in a baseball game. But I doubt that my grandfather knew this any more than I did. He meant we women were (or should be) the passive ones, responding to men's initiatives.

16 Peristiany (1966) describes this pan-Mediterranean concept, which applied rather well in Turkey in the 1950s.

17 Wall et al.'s (2018) recent research in Turkey suggests that much about gender relations remains as in the 1950s in rural Turkey.

18 On a brief visit to Turkey in 1990, my son (then nine), who'd lived the previous four years in Oman, was shocked to see Turkish men driving without shirts. Modesty was also important for *men* in Oman.

19 'Pussy' is an American slang term for a vagina and is also used to refer to a woman's sexual availability (perhaps now more commonly recognized, given President Trump's usage).

20 See Fine (2010) or Saini (2017) for evidence on the ineffectiveness of this strategy.

21 It was during our first professional job search that I encountered one of my first flagrant cases of gender bias: a male interviewer discovered that I also had a qualified husband

on the job market. He suggested that my husband take the job and I work as a secretary, hoping "something would turn up". When I gently pointed out that I was better qualified for their particular job than my husband, he responded that it would be too socially embarrassing to hire the wife and leave the husband unemployed (Colfer 1971).

22 Although the images from that forest after the fires of 2016, show areas of wasteland (https://en.wikipedia.org/wiki/San_Bernardino_National_Forest#Vegetation), prior to that the adjacent mountainous area sported many conifers: "ponderosa pine, Jeffrey pine, sugar pine, Coulter pine, lodgepole pine, single-leaf pinyon, and knobcone pine . . . white fir, bigcone Douglas-fir (*Pseudotsuga macrocarpa*), incense cedar, and western juniper. . . . Canyon live oak, California black oak, and Pacific dogwood also grow here. The forest contains an estimated 87,400 acres (354 km2) of old growth. The most common old-growth forest types are Sierra Nevada mixed conifer forests, white fir (*Abies concolor*) forests, Jeffrey pine (*Pinus jeffreyi*) forests, and lodgepole pine (*Pinus contorta*) forests" (https://en.wikipedia.org/wiki/San_Bernardino_National_Forest#Vegetation, accessed 29 May 2019).

23 Most of our time was spent in the arid, treeless south, but we initially spent some time in the forest-rich north. We only stayed in the country for five months as we both contracted hepatitis and had to return to the US.

24 Discussed in Colfer (1983), also available in Colfer et al. (2017).

25 "At the core of a companionate marriage is friendship and trust and the belief that both partners have equal responsibility in all domains of the marriage. They share the economic burdens and child rearing, and they believe that both partners' sexual needs and wishes should be clearly articulated and fulfilled" (Wallerstein and Blakeslee 1995, p. 155). Bernard (1982) describes it as stressing "affection, comradeship, democracy, and happiness of members of the family" (p. 126).

26 I focus here on the American instances, because they served to form my own views very fundamentally; the Middle Eastern experiences served as valuable counter-examples but were not central in my personal views of appropriate masculinities. I was not tempted to marry a Middle Eastern man.

27 Brandth and Haugen (2000, 1998) examine the significant changes in Norwegian masculinities as shown in forestry journals over time; Anderson (2009) looks at recent changes in the US and the UK, with regard to sports (both discussed further in Chapter 7).

28 My mother tells me in 2019 that my father willingly hung out the laundry when I was a child (1940s) but insisted on doing it at night so no one would know he was doing it.

References

Anderson, Eric. 2009. *Inclusive Masculinity*. New York: Routledge.

Bernard, Jesse. 1982. *The Future of Marriage (1982 Edition)*. New Haven, CT: Yale University Press.

Brandth, Berit, and Marit S. Haugen. 1998. "Breaking into a Masculine Discourse: Women and Farm Forestry." *European Society for Rural Sociology/Sociologia Ruralis* 38 (3): 427–442.

Brandth, Berit, and Marit S. Haugen. 2000. "From Lumberjack to Business Manager: Masculinity in the Norwegian Forestry Press." *Journal of Rural Studies* 16:343–355.

Colfer, Carol J. Pierce. 1971. "Woman in Anthropology." *Anthropology News* 12 (4):9.

Colfer, Carol J. Pierce. 1975. "Bureaucrats, Budgets, and the BIA: Segmentary Opposition in a Residential School." *Human Organization* 34 (2):149–156.

Colfer, Carol J. Pierce. 1983. "On Communication Among 'Unequals'." *International Journal of Intercultural Communication* 7:263–283.

Colfer, Carol J. Pierce, Marlène Elias, Bimbika Sijapati Basnett, and Susan Stevens Hummel, eds. 2017. *The Earthscan Reader on Gender and Forests*. London: Earthscan/Routledge & CIFOR.

Fine, Cordelia. 2010. *Delusions of Gender: How Our Minds, Society and Neurosexism Create Difference*. New York: W. W. Norton & Co.

Gibson, Chris. 2013. "The Global Cowboy: Rural Masculinities and Sexualities." In *Sexuality, Rurality, and Geography*, edited by Andrew Gorman-Murray, Barbara Pini, and Lia Bryant, [no page numbers]. Lanham, MD: Lexington Books.

Gibson, Chris. 2016. "How Clothing Design and Cultural Industries Refashioned Frontier Masculinities: A Historical Geography of Western Wear." *Gender, Place & Culture: A Journal of Feminist Geography* 23 (5):733–752. https://doi.org/10.1080/0966369X.2015.1058758.

Grindal, Bruce. 2011. "Confrontation, Understanding, and Friendship in a Redneck Culture." *Anthropology and Humanism* 36 (1):89–100.

Li, Tania. 2007. *The Will to Improve—Governmentality, Development, and the Practice of Politics*. Durham: Durham University Press.

McGinley, Ann C. 2004. "Masculinities at Work." *Oregon Law Review* 83 (2):359–434.

Peristiany, J. G., ed. 1966. *Honour and Shame: The Values of Mediterranean Society*. Chicago, IL: The University of Chicago Press.

Putz, Francis E., and N. Michele Holbrook. 1988. "Tropical Rainforest Images." In *People of the Tropical Rain Forest*, edited by Julie Sloane Denslow and Christine Padoch, 37–52. Berkeley, CA: University of California Press.

Saini, Angela. 2017. *Inferior: How Science Got Women Wrong—And the New Research That's Rewriting the Story*. Boston, MA: Beacon Press.

Veri, Maria J., and Rita Liberti. 2019. *Gridiron Gourmet: Gender and Food at the Football Tailgate*. Fayetteville, AR: The University of Arkansas Press.

Wall, Jeffrey R., Elif Başak Aksoy, Nesibe Köse, Taner Okan, and Coşkun Köse. 2018. "What Women Know that Men Do Not about Chestnut Trees in Turkey: A Method of Hearing Muted Knowledge." *Journal of Ethnobiology* 38 (1):138–154.

Wallerstein, Judith S., and Sandra Blakeslee. 1995. *The Good Marriage. How and Why Love Lasts*. Boston, MA: Houghton Mifflin Company.

Wignall, Ross. 2016. "From Big Man to Whole Man: Making Moral Masculinities at the YMCA." In *Masculinities Under Neoliberalism*, edited by Andrea Cornwall, Frank G. Karioris, and Nancy Lindisfarne, 183–197. London: Zed Books.

3

IMMERSION IN RURAL AMERICA

The case of Bushler Bay[1]

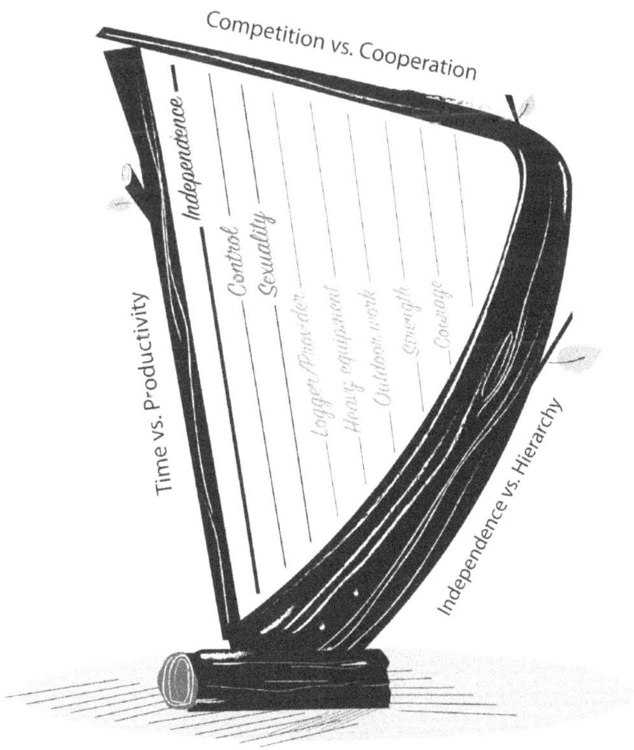

A Bushler Bay Logger Harp Example

Introduction and reflection

In late 1972, my then-husband, Michael, and I moved to Bushler Bay, Washington, where we were hired as 'on-site researchers' by Abt Associates.[2] The need to care for our young daughter (then three), our near-final stage in obtaining our doctorates and our interest in coming up with more equitable gender roles led us to suggest job-sharing to our future employers. We argued that this would grant them the expertise of two people and the likelihood of >50% time investment from each of us for the same salary and benefits. We convinced them that this was a sufficiently 'cutting-edge' idea for them to try it out.

Our doctoral programmes at the University of Washington had not prepared us for research in rural America—we'd expected to work in the Middle East—so we weren't at all sure what we would find in the two research communities. We arrived in the smaller community mid-winter, having just bought a Dodge pickup truck, partly as a potential rapport builder with the loggers we imagined we would (and did) find and partly because of Michael's rural American, masculine fascination with trucks.[3]

At the time, I was newly aware of the degree to which women were disadvantaged professionally. I'd begun going to 'consciousness raising groups' and 'assertiveness training' to address gender inequities personally; Michael supported me in these efforts.

In those days, anthropologists were expected to avoid influencing local cultures, and we struggled with this dictate, given our concerns about gender roles and sexual politics in general. Our decision: To avoid any 'proselytizing' but to be honest about our own preferences and practices, which were quite strange to our neighbours. We recognized that our knowledge of English, shared ethnicity[4] and familiarity with American culture generally were advantages, but that we would also have to struggle to minimize our own cultural assumptions and biases as we viewed local realities.

Anthropologists tend to be drawn to cultures that differ from their own; it's a central reason many study anthropology. We are intrigued by 'the other'. In this rural American context, there were two groups, described later, one of which (the 'Locals'),[5] more different from my own background, aroused my curiosity. Although I recognized that Local views were more antithetical to feminism, which I embraced enthusiastically, I still found their independence and disdain for middle-class conformity appealing. Recognizing this conflict, I also found Local men's more overt sexuality attractive.

One version—of many possibilities—of a logger's common harp is reflected in the image under this chapter's heading. The differing shadings of the strings reflect clustering and importance, with the thicker strings more commonly admired within the value system represented. To put our findings in context, we need to reflect a bit on the environmental setting in which these tones and clusters are selected and 'songs' created.

Forests and the environment

Bushler Bay, comprising two unincorporated communities (one of ~550 people, the other ~250 (Colfer, Cerveny, and Hummel 2019), sits on the Eastern shore of the Olympic Peninsula, bordering Hood Canal (in Puget Sound; Figure 3.1). The area was (and remains) sparsely populated, with dense rainforests to the west and south. To the flatter north, though still forested, there were more agricultural fields interspersed with forested area. Bushler Bay's economy was largely dependent on logging throughout the 20th century until national concerns about the spotted owl (*Strix occidentalis*) and the marbled murrelet (*Brachyramphus marmoratus*) spawned the 'Timber Wars' (Colfer 2018), eventually closing down much of the timber production[6] on the east side of the Peninsula. Families routinely engaged in hunting, fishing and collection of non-timber forest products, like mushrooms, a wide variety of berries and firewood, with several small industries collecting brush, particularly salal (*Gaultheria shallon*).[7]

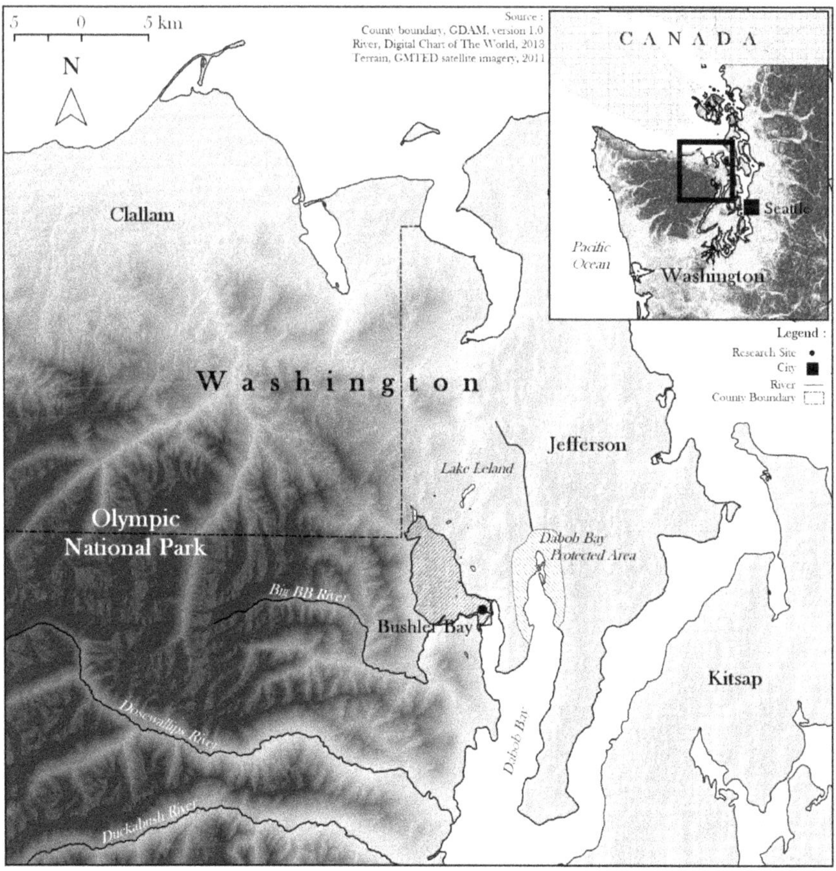

FIGURE 3.1 Map of Bushler Bay and surroundings.

Topography is steep and covered with dense stands of Douglas fir (*Pseudotsuga menziesii*), cedar (*Callitropsis nootkatensis*) and hemlock (*Tsuga heterophylla*), Washington's state tree. Land ownership and management represented a mosaic, including large- and small-scale private ownership, but most lands on the Olympic Peninsula were (and remain) managed as part of the Olympic National Forest and National Park.

Formal forest managers from the US Forest Service were very evident in day-to-day life, intimately involved in managing this valuable forest for sustainable timber production, most obviously in taking bids and letting contracts for timber sales to loggers. Most policies were crafted in an explicitly top-down manner in distant Washington, DC. I remember one community meeting called to record the views of locals on several management options (including the creation of an inaccessible wilderness area), without apparent influence on ultimate decisions taken. Otherwise, government regulations were followed and enforced by USFS personnel. Large-scale private management was fairly similar at that time, with managers also obtaining bids and letting contracts to local loggers. Smaller-scale owners had comparative freedom to manage as they chose.

Pingree and DeLuca's (2018) study just south of Bushler Bay provides a nice introduction to the area, which falls:

> within the North Pacific maritime dry mesic *T. heterophylla/P. menziesii* potential vegetation zone, between 400 and 600 m in elevation . . . Soils of the northern and eastern Olympic Peninsula are formed on marine deposits of basalt, colluvium, volcanic ash, and glacial till.
>
> *(p. 97)*

Soils were classified within the US Soil Taxonomy classification as Haploxerepts (NRCS, 2014):

> A volcanic ash layer was present at [two] sites . . . Mean annual temperature from 1980 to 2010 averaged 8.1 ± 0.5°C and mean annual precipitation was 2181 ± 592 mm across all sites . . . . The Hargreaves climate moisture deficit ranged from 145–291 across all sites and is reported as the sum of the monthly difference between the potential and actual moisture deficit
>
> *(p. 97)*

Burwell (n.d.) describes the history of fire danger, which was minimal until this century, when fires have increased, partially because of changed management practices and drier conditions.

In the following discussion, I return to the 'cultural harp' analogy raised in Chapter 1. As noted earlier, the harp itself represents the comparative stability of cultures. In these two communities, we might consider the harp[8] to have its usual three sides: in this context, one relates to people's perceptions of time, the second to a balance between competition and cooperation and the third deals with the balance between independence and hierarchy. These are alluded to in the following sections, which place the strings on this harp in cultural perspective. Men from these communities (and to some extent, women as well) have the option of 'plucking' one or another of these strings at any given moment. Each man can create his own 'chords' and 'song' based on the strings he opts to pluck. Recognizing this freedom allows us to capture the variety and dynamism of gendered ideas and behaviour.

Masculinities and the provider harp string

On arrival, A.M. Colfer and I quickly became immersed in the planning process for what was called locally 'the federal project'. The project was led by one of the school teachers, with help from several others; an important component of the project was to mould the curriculum in such a way as to make it more welcoming for local youth, particularly boys, among whom the motivation to succeed in school was minimal; the boys' dropout rate was high (as in Kalimantan, Chapter 4). There were several older community members who participated actively, helping to craft the proposal and begin implementation. Almost all the planners were men, most middle aged or older; the key decision-makers were all men.

Very quickly we learned one of the central features (or notes on the harp) of masculinity throughout the community: men were seen as the breadwinners.[9] Indeed, this was one of the few elements of masculinity shared between community groups—though the manifestation of breadwinning differed. Coming into the community, Michael and I presented ourselves as equally involved in the research project, equally employed. Yet whenever I was introduced to someone new, this got transmuted into my being "married to Michael, who works for the federal project". This perception continued throughout the three years we lived there, despite our frequent corrections.

It wasn't as though no women worked. Bushler Bay School represented the only real cluster of employed women in 1971–1972, including 15 women and 16 men.[10] But careful analysis of these figures revealed several patterns that reinforced and reflected both men's expected roles as breadwinners and their greater importance in work settings: First, men dominated the higher status, certified (teaching) staff: 13 men, 4 women; and women dominated the non-certified (kitchen, maintenance) staff: 3 men, 11 women. Second, men's jobs were treated as important, women's not so much.

- Men held all administrative positions (superintendent, principals, head of the federal project) and taught the higher status 'men's subjects' (math, science, social studies) as well as woodshop and coaching; women taught little children or 'women's subjects' (English, writing, reading).
- Men predominated in high school, considered to require greater expertise, while women taught the lower grades.
- Men tended to have clearly defined contracts with reasonable job security, whereas women were often hired from year to year (not knowing, for instance, during the summer if they'd be employed in the fall).

- Negotiations over wages or promotion were private for men; public (conducted during school board meetings), embarrassing and demonstrably less successful for women.

Age, another social structural principle in the community, intersected with gender. Almost all men who taught or administered in school were in their 30s and 40s, their 'prime of life', enmeshed in their masculine role as breadwinner; women were mostly represented at the extremes: single and in their 20s or in their 50s. This school context, in which boys (and girls) were locked for six hours a day, 280 days a year, for 12 years, provided a powerful statement about what the broader society expected of men (and women) with different qualifications and at different ages, and what the associated benefits were.

Overall, the community sentiment was that men needed their jobs and that women worked for 'extras', even when the evidence suggested otherwise.[11] Many men, particularly those long-resident in the community (Locals, discussed below), tended to object to their wives working, considering it an affront to their ability to support the family. This was a somewhat moot point, with so few employment opportunities for women. Women also expressed their reluctance to have a job that paid more than their husband earned.[12] There was some evidence that periods of men's unemployment (common among Locals) resulted in more marital discord. Certainly, that was women's perception.

Masculinities and the sports harp string

Besides the school as a work setting providing a daily reminder of culturally appropriate roles for men and women,[13] one of its key functions was to organize sports events. Such events were the most visible form of entertainment for the community, especially during the bleak winter months.[14] A high school basketball game was colourful, active, entertaining and watched and appreciated by all sectors of the community. It was also a vivid portrayal of gender roles.[15] The team members, boys, were chosen for their strength and athletic prowess. Cheerleaders were selected based on beauty and popularity.[16] Little boys idolized the players and little girls longed to become cheerleaders.

A basketball game was rife with symbolism. The team members shared one uniform, cheerleaders another. As the national anthem played to signal that the game was about to begin, the two opposing teams were arranged diagonally across the court, players in the centre, with their respective cheerleaders on the ends of the diagonal. Players—the central stars of this event—were segregated from the audience by their seating, their access to the courts and their sole access to the locker room. Cheerleaders also had their own space, standing together near one basket. They took over the court whenever the boys vacated. They provided snacks for the team and for audiences at home games. Their primary role was to instill 'school spirit' and stimulate audience support for the team, which was expressed by loud cheering.[17] Their cheers both urged the team on and reassured them when they failed.

The team's role was to win games. They did this by striving for cooperation, organization and teamwork in their play, within a context of hierarchy where

> 'men control other men' . . . [T]he referees decide when a foul or transgression has occurred, and players comply; the principal reminds people not to walk on the floor, and by and large they do not; the coach directs the team's play; and the captain of the team makes decisions about the team's strategy and time-outs.
>
> *(Colfer and Colfer 1979, p. 191)*

The 'hidden curriculum' of a game demonstrated expected livelihood roles for men—to work with other men within hierarchical contexts, where groups competed against each other.[18] But it also portrayed the marital roles for men and women.[19] What mattered in a game was who won. The girls could provide emotional support, but they could never score a point, just as most Bushler Bay wives could not contribute to household income.[20] The importance of this curriculum was manifest in wide attendance at games, periodic community controversies about coaches, a preoccupation with the behaviour of players (and cheerleaders), community interest in the events and a general concern about the win-loss records of the team.

In addition to these recurrent representations of gender roles at games, the importance of school sports was solidified every spring when an election to pass the 'school levy' was held. The levy was needed to supplement the funds available to run the school, and every year, cutting the sports programme was floated as a possible outcome should the levy fail. This issue was passionately debated in the community, with some men particularly adamant about the importance of sports for boys. All recognized and valued its role as community entertainment. But there was a more visceral attachment to it among one community group, Local men. They argued that sports taught boys to be men, to compete (as they would have to do in the world of work),[21] to cooperate as a team, to learn leadership and to strengthen their bodies physically.

Cooperation and competition were concepts that came up routinely. When Locals talked about young boys learning to be men, they stressed the significance of competition. They saw a competitive world 'out there' in which boys would need to strive to excel. Yet interestingly, local work patterns were remarkably cooperative. The danger of their work meant they relied on each other for their very lives. They were not in fact competing on a particular job; each had a specific role at a given time, and they were complementing each other's actions.

Among Public Employees, the emphasis at work was explicitly on cooperation within a given unit, competition reserved for other units (for funding, for personnel, for equipment, in segmentary opposition; Colfer 1975). This was clearly taught in the school, where one grade would cooperate internally to compete against another, girls would work together to win out over boys, who were likewise cooperating among themselves.

Competition, so closely tied to masculinity for Locals, was conceived as cooperation's opposite. Men were seen to be competitive and women cooperative; yet clearly in real life, whether Local or Public Employee, man or woman, both competed and cooperated. We came to see it as a cultural balancing act.

Two ideals of masculinity

The divide between Locals and Public Employees only gradually dawned on us (Colfer and Colfer 1978), although differences in dress should have tipped us off.[22] Logging was the preeminent exemplar of the Local way of life, and loggers had a distinctive dress: [often dirty] baggy jeans, red Logger's World suspenders and caulk boots. Different public institutions required different dress, with US Forest Service personnel, the stereotypical exemplar of the Public Employee lifestyle, wearing green uniforms. In institutions without uniforms, Public Employees tended to wear neat slacks with collared shirts and sometimes ties. In general, Locals dressed less formally than Public Employees.[23]

School personnel and other Public Employees spoke of this Local-Public Employee differentiation as one of social class.[24] But we found viewing it as cultural difference made more sense. Those Public Employees labelled 'lower class' ('Locals') included a great variety of qualities incompatible with that classification.[25] There were wealthy people, educated people, people of longstanding residence, mixed in with others who might be widely termed 'lower class', all sharing identifiable and shared predilections.

The more I have thought about 'class' in the US, the more convinced I have become that such an appellation effectively reinforces and replicates the existing power structure, rather than clarifying differences among groups. Those with 'middle-class' values define what constitutes prestige and value. And they have a full arsenal of weapons, not least of which is control of the school systems—to which all American children go every day. Here, I use a variety of kinds of encounters to present the evidence of different masculinities—different selections of 'notes' on the harp—among these two groups.[26]

The variation in the use and perceptions of time between the two groups makes a nice segue from the presentation of masculinity in the Public-Employee-dominated school context. In school, as in the bureaucracies where Public Employee men worked,[27] time structured life. Students and employees had to be at work on time, events and meetings were set for specific times, deadlines were set and punctuality was both expected and rewarded. Being on time and meeting deadlines (among other things) were evidence of good job performance. Success within bureaucracies was also not unrelated to length of time on the job. My roughly 24 years in the American educational systems by then prepared me well for this orientation to time.

Local men, however, enmeshed as they were in non-bureaucratic settings—logging crews, oyster culture, family businesses—tended to see following timetables as irrelevant. They evaluated their own work based on production (number of logs to the landing) and income earned. Time was mainly relevant insofar as it could indicate efficiency.

The link with masculinities, typically contrasted in this setting with femininity, was strengthened by the fact that both Local and Public Employee women—as primary caretakers of their children—remained enmeshed in the school's somewhat arbitrary timetables. Additionally, the efficient accomplishment of women's usual tasks was difficult given the necessarily responsive nature of caretaking. Cooking dinner or doing the laundry could not always be done efficiently: a child might need his diaper changed or an elder have a medical emergency. Nor did it behoove women to assess the value of their efforts by income earned. Gathering, gardening, canning, handicrafts—the ways women without formal jobs typically spent their time—were notoriously unprofitable in terms of income. Women more often assessed their work in terms of time spent rather than either amount produced or money made.

Local men's approach to time became a feature of their masculinity, strengthening their cultural conflict with Public Employees and with women—multivalent conflicts, as shown next.

Masculinities and the control harp string

The control harp string begins with men's control of their own emotions. As with the men in my own families (Chapter 2), Bushler Bay's men, neither Locals nor Public Employees, cried in public or to my knowledge, in private.[28] They were unlikely to discuss their emotions, except in the most intimate of circumstances and many avoided the topic even then.

Local men were more overt about demonstrating control over women than Public Employee men (though the school's 'hidden curriculum', described earlier, shows Public Employee men's dominance as well—just more subtly). Some Local men believed they had the right as men to 'rule the roost' at home. That might mean ordering a wife to 'get me a beer', 'go buy cigarettes', 'shut those kids up', etc. My own suspicion is that when the 'breadwinner string' on the cultural harp was out of kilter, the 'ruling the roost' string was more likely to be twanged.[29] Jokes about a man as 'pussy whupped' (ruled by his wife, unable to resist her sexuality and exert manly control), for instance, were a common and humorous way men could differentiate themselves from a man so marginalized and from women in general. Accusing someone else of this 'failing' also reinforced one's own position as in control at home.

In both Local and Public Employee households there tended to be a simple expectation that of course a man's wife would cook dinner, maintain a clean house, do the laundry and take care of the children. The commitment of Public Employee couples to replicating an idealized form of middle-class life—nice house and car, mowed lawn, more formal and stylish clothing—meant a greater interest in having these tasks performed well than was found among Locals. This may have resulted in stronger if more subtle pressure on the Public Employee wife to keep a cleaner house, prepare more complex meals, etc. The greater commitment to 'togetherness' for Public Employee married couples also meant such wives were subject to more husbandly monitoring of their behaviour than were Local wives.

The response to Michael's and my [locally perceived] peculiar division of labour—alternating childcare daily, sharing domestic tasks evenly—differed between groups. Public Employees clearly found it odd but did not, to my knowledge, criticize. Some men would admit to sharing some of the childcare but drew the line at changing a diaper or doing the dishes, tasks that carried extra symbolic weight as unmanly.

Among some Locals, however, our division of labour was cause for utter disdain. I remember sitting with an older Local couple at a party. Michael said something in passing about what he'd cooked for dinner. The wife gave him a withering look and asked sarcastically if he wore a frilly apron when he cooked.[30] She was clearly challenging his masculinity and expressing her disapproval. I was told that some Locals considered me a 'ball buster' (a dominating woman who destroys a man's masculinity), partly because of my feminist leanings but also probably because of Michael's willingness to share domestic tasks.

Another occasion in which my purported 'ball busting' came to the fore was at the Whistling Oyster Tavern, the only night-time recreational setting in town and one where Locals (unlike Public Employees) tended to gather. The tavern provided alcohol, music and several games like foosball and billiards. This particular winter evening, I was on a roll and, unusually, doing very well at the various games. Indeed, I kept on winning. Eventually, two women, separately, came to me with friendly warnings: "Quit winning", "Let the men win". They worried that the men would be upset. I sensed a worry about violence somehow—though I was never entirely clear what they thought the men might do. Or perhaps they feared that continued winning could worsen my reputation as a 'ball buster'. I never felt in danger nor did I see or hear of evidence of physical abuse (though it probably sometimes happened). I suspect these warnings were prompted by the women's sense of responsibility for men's emotional health, which included the men's security in their own masculinity (doubly endangered when unemployed, so common in winter). Robinson and Hockey (2011) found a similar idea in northern England:

> Women's interviews showed that a capacity for empathic warmth, a commitment to emotional expressivity, and a felt concern at the prospect of any threat to a man's emotional well-being were seen, by women, and men, as desirable and indeed necessary qualities that were intrinsic to femininity.
>
> *(p. 170)*

Women reinforce these cultural notions of masculinity and femininity as well as men.

Masculinities and the strength and courage harp strings

Local men, in reference to their own masculinity, first stressed physical strength and dangerous, outdoor work. The simple fact of working outdoors implied an ability to withstand difficult weather (rain, snow, heat), and it typically included an appreciation for the beauty of the woods. Many loggers found painful the environmentalists' common accusation that loggers were destroyers; most saw themselves as providing a valuable product for society in an [under-appreciated] efficient and sustainable manner.

Considerable physical strength and skill are necessary to handle a commercial chainsaw[31] and the other heavy equipment used in logging. Metropolitan Life Insurance Company rated Pacific Coast lumbermen as the worst risk for 'death by violent accident' (Newsweek 1974),[32] and everyone in town knew people who'd been killed or maimed by a falling tree or upturned log truck. Climbing up a tree to cut off the top was among the most dangerous. Reed (2003) documents similar, more recent dangers in logging on nearby Vancouver Island; Carroll (1985) describes Oregon loggers' pride in both their levels of skill and their courage[33] in confronting danger. The manly occupation of logging is routinely contrasted with bureaucrats' 'paper-shuffling' and with women's work.

The words on a log truck driver's T-shirt in Oregon sum up a common Local view: "If you ain't a logger, you ain't shit", a sentiment expressed more genteelly in Carroll (1985). This combined the idealization of logging with some important interactional qualities (lack of regard for 'proper' grammar, straight speaking and easy use of profanity), which served also to differentiate Locals from Public Employees and women.

Local men eschew words like 'please', 'thank you', 'do you mind?'. A. Michael Colfer's 1974 fieldnotes capture this attitude:

> We three stood at the end of the bar picking up beer for a party. The young logger was hosting at his home, and we were getting supplies at the tavern and the store. The three of us were laughing and enjoying ourselves. The Public Employee (not, incidentally, a Forest Service employee) tried for the third time to pay for part of the beer. Again, the logger refused.
>
> "Shit, man, I'll pay for my part".
> "Nah, I don't want your dirty ol' money".
> "Take it. Please".
> The word was like a red flag.
> "Don't give me that Forest Service shit!"
> (Colfer and Colfer 1979, p. 45)

Carroll (1985) emphasized Oregon loggers' commitment to their own independence. As *men*, they wanted to be able to determine their own lives, not to be subject to a boss's dictates, and have the freedom to quit any job they did not like.[34]

These were sentiments reiterated often in Bushler Bay, where at that time logging—requiring both strength and courage—was still a successful livelihood strategy. Independence, manhood and logging were intimately connected for Locals.[35]

BOX 3.1 A LOGGER'S CHORD—BREADWINNING, INDEPENDENCE AND DIFFERENTIATION FROM WOMEN

The breadwinner harp string sometimes conflicted with the Local independence string, when for instance a logging boss exerted his supposed authority, or a co-worker questioned a man's competence. The manly thing to do in such circumstances was to quit that job and seek another in the woods. The sexuality harp string had an impact as well: Other loggers admired a man's independence, whereas his wife might value his breadwinning more. In the US at that time many saw gender relations as a 'battle of the sexes', in which all men were pitted against all women. A man who paid attention to his wife's concerns about his breadwinning was in danger of being considered 'pussy whupped'. The greater adherence among Public Employees to the breadwinner role and to related security put these officials closer to the category of women and earned them a reputation as 'paper pushing lackeys of the bureaucracy'.[36]

This enthusiasm for independence was rarely expressed by Public Employees, who recognized and tended to accept their dependence on their bureaucratic hierarchy. They traded independence on the job—a Local choice of tone on the harp—for security and reliable benefits, which they linked to the reliable breadwinner string. Although Public Employees also complained about their bureaucracies, many saw value in working within one. They were part of a cadre of people with common goals, they were (distantly) linked to the nation's political and decision-making centre, and some appreciated the simplicity of having someone else making day-to-day decisions about the work to be done. Their jobs in a bureaucracy allowed them to pursue the 'middle-class' lifestyle they sought. Bureaucracy and hierarchy did not appear to link directly or negatively with masculinity (also seen in Sitiung, Chapter 5).

As noted earlier, having an occupation was a crucial element of men's identity, a particularly important tone on their harp, for both Locals and Public Employees. But among Public Employees their profession was more closely linked to the breadwinner tone than for loggers (who typically had periods of unemployment every year). The latter saw their profession as intrinsically valuable and good, a source of pride closely linked, via strength and courage, to manhood.

Masculinities and sexuality

Altork (1995), an anthropologist, in examining her own behaviour and attitudes toward sexuality in the field (in neighbouring rural Idaho), recounts her attraction to the firefighters she studied. This quotation from her writing captures both the inherent sexuality of this dangerous outdoor work (within the rural northwestern American context) and the ambivalent attractiveness of it to (American) women—a response I also experienced. Altork had asked to photograph two men friends out on firefighting duty; one had suggested they clean up first. The other responded:

> 'What do you mean? This is how we look out there, man! This is what it's all about—men and machines, sweat and grit. Why wash it off?' Turning to me, his face opening into the grin of the heartbreaker, he adds, 'Hey, Katie, this is the way it really is, right? It's hot and heavy work. But, what the heck, it's kind of sexy, don't you think?' And then, looking me full in the face, he winked. And I winked back.
>
> *(p. 133)*

In Bushler Bay, Local and Public Employee attitudes about sexuality varied, particularly in how public it was—and thus how much access I had to people's views about it. Public Employees were unlikely to talk about sex in public, discuss marital infidelity openly or act in sexually demonstrative ways. Their dancing was usually restrained; any marital infidelity was carried out with utmost discretion. Public Employees tended to disapprove publicly of non-marital sexuality at any age.

Locals were much more open about the topic. Local men might brag about their 'conquests', and some women readily admitted to such adventures as well. Among both genders, many considered infidelity more of a peccadillo than a serious offense. Sexually active Local girls and women, unlike boys or men, endured some disapproval, but Local attitudes were much more forgiving of female sexuality than were Public Employees'. Sexuality among Locals was ultimately considered a personal decision for both sexes. Public Employees expressed disapproval of this attitude, and stronger disapproval of overt female sexuality. Pascoe (2007) describes the subtler but pervasive reinforcement by school personnel of heteronormativity in the California school she studied, which of course implicitly requires female sexuality.

Sexuality is related to gender everywhere, but its relevance was very overt among Local men. In day to day discourse, sexual innuendo was common: A log truck driver would refer to his logging rig as 'my Peter'.[37] A bumper sticker advertised "Old loggers never die; they just get a new Peterbilt."[38] The symbolic relationship in Local men's minds between large logs, powerful machines, strong men and sexuality was clear.[39]

Women appreciated and reinforced this link. Young girls were drawn to those boys who excelled in sports or other physically demanding tasks.[40] This excerpt from A.M. Colfer's 1973 field notes captures a common feeling:

> We sat in her living room talking about the school, local life, and logging. At one point she got up and went into the other room, bringing back with her a blown-up photograph of her husband in hard hat and overalls standing beside the log that he had just cut down—it filled the log truck.

> "You can talk about the romance of various occupations all you want," she said, "but it's going to be pretty hard to beat that". She looked up, smiled proudly, and added, "Now there's a man!"
>
> *(Colfer and Colfer 1979, p. 39)*

Research by Reed (2003) in nearby Canada showed that

> women who identified themselves within 'traditional' feminine relationships such as helpmates and wives as well as women who self-identified as participants and partners in forestry occupations suggested that feminine identity in forestry communities were [sic] locally and mutually constituted with those of men's masculine identities.
>
> *(p. 384)*

Adultery was common in the community, as was marital instability. Relations between many Local husbands and wives were full of suspicion. The atmosphere at Local parties, which typically involved considerable alcohol consumption, was one of sexual electricity. Men and women interacted with people other than their spouses, danced with each other (often with explicit sexual moves), argued with each other and some might disappear into the nearby forest for an illicit rendezvous. Many Locals had known each other since childhood, with their romantic histories and associated potential jealousies known and remembered. Unlike my father's attitude (that he should violently protect my mother from unwanted attentions), I never witnessed a fight between men over women. Men tended to withdraw in the face of wifely flirtations with others. And unwanted attentions were gently and effectively discouraged more subtly, by laughter, words and/or meaningful looks from friends, both men and women.

Public Employees tended to gather in smaller groups, segregated from Locals, with the emphasis on couple to couple interaction, rather than people operating socially as individuals. Any sexual attraction was kept strictly under control in public.

Although Local women complained about their husbands' infidelity, one concluded "but, you know, even though I have to put up with a lot of shit, I know he's a real man. I wouldn't want to trade that" (Colfer and Colfer 1979, p. 39).

In mid-1977, after I'd left the community, I conducted an anonymous survey. Among other things, I asked a series of questions about the various birth control options available. To my great surprise, 23% of the 26 respondents' husbands had had a vasectomy (Colfer 1977, p. 51). In the three years of my fieldwork, only one Public Employee wife, my closest friend there, had told me about her husband's vasectomy. I'd heard several men say that they'd never have one and that birth control was the woman's responsibility.[41] Vasectomies were explicitly seen as attacks on their manhood (as expressed by my maternal grandfather), at least publicly. I have no way of knowing what the proportion of Local vis-à-vis Public Employee respondents was. But if there was a significant number of Locals in the group, the fact that the women kept their husbands' secret was interesting in itself—a strong statement on the depth of feeling among men (and perhaps related fears among women, who may not have wanted their husband's masculinity questioned).

Dominant masculinities summarized

Ideology, men's characteristics and interests and gender roles are some of the ways masculinities are framed and practised. Below I highlight those most consistently and publicly visible in Bushler Bay. As noted in Chapter 1, men in all cultures have some choice about the harp strings they pluck, constrained by the harp(s) of their contexts.

Ideology

The Bushler Bay community manifested a clear ideology of difference between men and women, with an expectation of men's dominance as something 'natural', related to men's and women's inherent natures.

Characteristics

Men were expected, particularly among Locals, to be strong, courageous, competitive, independent and dominant within the home. Among Public Employees, strength and courage were admired. Cooperation within one's bureaucratic work unit was emphasized, with competition reserved for other units, other sectors. Dominance within the home was more implicit, assumed, than among Locals.

Interests

Heavy equipment, sports, sexuality and being outdoors were key interests for many men of both groups, though more intense for Locals. Public Employee men tended also to be concerned with fulfilling middle-class expectations (nice home and car, secure employment, 'proper' clothing). Men in both groups showed an interest in differentiating themselves from the other group.

Gender roles

Being a breadwinner was an expectation for both Local and Public Employee men, though it was more central for Public Employees generally than for Locals. Although most men were fathers, this role was not stressed publicly (parenting often left largely to wives).

Bushler Bay men in forests: conclusions

There was a 'division of labour' in Bushler Bay, with many Locals directly dependent on the forest and its harvest for their livelihoods. Public Employee men, unlike the loggers, were intimately involved in national and state natural resource management hierarchies, particularly related to forests, via their roles as breadwinners. Politics at the local level were far more the domain of women (Colfer 1977)—so different from the Indonesian contexts to follow. Clubs provided most local services (streetlights, fire protection, entertainment, adult education) performed elsewhere by formal governance structures. Women were seen to have more time for such activities. Volunteer firemen were at that time all men, which fit both with the availability of men trained in fire suppression from the US Forest Service—though the environment was so wet that forest fires were rare and small—and with experience and appreciation of heavy equipment.

The level of conflict increased steadily during this period, as the federal government expanded its involvement in forest management, from timber harvest and replanting to attention to biodiversity losses. Pressures from environmental groups, trying to save the spotted owl and marbled murrelet, were another key factor, with the beginnings of adverse effects on Local livelihoods, and ultimately, on US Forest Service personnel with an interest in remaining in Bushler Bay.

Much forest-related masculine humor in the 1970s was dark and revolved around the spotted owl. It reflected the real dangers to people's livelihoods and specifically to men's employment in the woods, so central to loggers' self-image: Printed on a T-shirt, 'Save a logger, kill a spotted owl', or the offering of 'spotted owl soup' on the menu at the local cafe. Laughter spurred by these jokes identified an individual as opposed to environmentalists, while simultaneously serving to identify him as on the side of loggers in local resource conflicts (see Meyer's (2000) discussion of the unifying and divisive functions of humor).

The identities and livelihoods of Local men (and their families), so bound up in the 'song' of physical, outdoor activity focused on producing timber, were in serious danger (see Chapter 7 for a 2017 update). There were few obvious alternatives that required those skills and interests. Public Employee men employed in natural resource management were also endangered as budgets were cut and national attitudes toward foresters deteriorated. Both loggers and foresters came to be seen by many as 'forest destroyers', with the latter seen as responsible for unsustainable management. The definition of sustainability changed from simply sustaining timber harvests to sustaining the far more complex forest and its non-human inhabitants.

The people of Bushler Bay, like the Kenyah of Long Segar (Chapter 4), were suffering from policies crafted in distant urban centres over which the local community had no significant influence.

In 1975, we left our home in Bushler Bay. I was tired of small-town living, of my neighbours knowing my every move and of struggling with the conceptual quandaries of what to share of myself and how to interpret my observations and insights about the communities. A three-year interlude in Seattle followed during which I continued periodic fieldwork, analyzing data from the communities and working with friends to start a non-profit company (PACT, Professional Anthropologists Consulting Team). Michael and I sailed to Hawaii in 1978, where I began a second master's degree in international public health. The concluding part of that degree programme was a field study in a developing country. I chose Indonesia because of my earlier problems gaining permission to do research in the Middle East. The University of Hawaii School of Public Health had close ties with that country. In June 1979, I went to Bali to conduct my field study.

Notes

1 This chapter draws heavily on the following sources, as well as fieldnotes, personal journals and my own memories (Colfer and Colfer 1979, 1978; Colfer 1977, 1978, 1983).
2 Our work was to conduct ethnographic research in two communities, which I lump as Bushler Bay (Colfer and Colfer 1979). We were also to collaborate with Abt Associates' quantitative, cross-site research team from Cambridge, Massachusetts.
3 This vehicle involved a negotiation in which I argued for a car with good gas mileage and he, admitting that he'd always wanted a truck, argued for it as a rapport builder. He won.
4 The local population was almost exclusively white.
5 When the term Locals is capitalized, I refer to a social structural/cultural categorization.
6 In 1988, the Olympic National Forest harvested 264 million board feet of wood; by 1994 the harvest had declined to 8 million board feet (Headwaters Economics 2012).
7 Ballard and Huntsinger (2006) look at local knowledge of this product on the Peninsula.
8 We could also look at a lower scale and imagine separate harps for each subgroup, analogous to ethnic differences in Chapters 4 and 5.
9 This study may also contribute to Westberg and Powell's (2015) call for "scholarship that focuses on how femininity and masculinity are created, providing a critical analysis of the production and reproduction of gendered norms in bureaucratic settings" (p. 1235).
10 Most data about the school come from Colfer (1978).
11 Careful analysis of cases revealed that those women who genuinely needed the money fared worst of all within the school system. About half of all marriages at the time ended in divorce; there were plenty of women who needed to make a living.
12 Reed (2003) reports that "some women who were interviewed [in nearby Vancouver Island] chose not to risk challenging the 'manliness' of their partner by taking a job" (p. 384). Women's sense of responsibility for the related emotional health of husbands is also reflected in Robinson and Hockey's (2011) study of masculinity in northern England.
13 Pascoe (2007), who studied a California school, notes, "The ordering of sexuality from elementary school through high school is inseparable from the institutional ordering of gendered identities" (pp. 26–27).
14 Temperatures range on average from 31°F to 45°F in January, with 54 inches of rain and many windy and cloudy days (www.worldclimate.com/climate/us/washington/quilcene, accessed 7 November 2018).
15 See Anderson (2009) for an analysis of the traditional role of sports in supporting hegemonic masculinities and in differentiating valued heterosexual men from disvalued women and homosexual men in the US and UK.
16 Being a high school cheerleader was valued far more than being on girls' sports teams. Boys' sports were ubiquitous and much better supported than girls' sports (Colfer and Colfer 1979).
17 Blanchard's (1974) analysis of boys' and men's basketball games in New Mexico compares Navajo and Anglo Mormon approaches and values. The Anglos approached the game seriously and more similarly to the Bushler Bay community. The Navajo men,

despite a cavalier attitude toward winning, called themselves the 'Warriors in Rimrock', while the Anglo team, whose interest depended on winning, called themselves, oddly, the 'Onions'.

18 This combination of cooperation and hierarchy fit more smoothly with one segment of the community (Public Employees) than the other (Locals)—a differentiation discussed below.

19 Pascoe (2007) describes the California cheerleaders as "working as football players' perky heterosexual helpmates" (p. 118).

20 Anderson (2009) notes that

> Women, adult men, and other marginalized boys pay tribute to elite men by supporting them in the very arena in which they struggle to maximize their influence–athletic competitions. The epitome of this is when women cheer for male athletes, relegating themselves to symbolic subservience.
>
> *(pp. 44–45)*

21 Also considered important in Jamaica (Tantam 2016): The football field "was a space for men to express themselves and compete over what it meant to be a successful man" (p. 222).

22 Locally competing masculinities are not unusual:

> They [Bristol-Rhys and Osella] describe how the deployment of racial stereotypes by indigenous Emirati and migrant Asians centre on competing masculine ideals, infused with coloniality: for the Emirati, Indians are neutered, cast as effeminate non-men; for Indians, a reified notion of the male breadwinner becomes the measure against which the Emirati are found lacking.
>
> *(Cornwall 2016, p. 17)*

23 Robinson and Hockey (2011) note in their study of masculinity in the UK:

> In the statement 'I'm a suit man', man and suit are elided in a business world where men can be referred to simply as 'suits', one item of clothing standing for their entire occupational identity.
>
> *(p. 104)*

See also Brandth and Haugen (2000) on the significance of attire in Norway's forests. Symbolic dress is also evident in upstate New York in 2020; professors at Cornell University wear slacks and a collared shirt; 'working men' visible at the diners around Ithaca's periphery are likely to wear jeans, T-shirts and work boots.

24 Colfer and Colfer (1979) quote a faculty member in 1974: "I tell you, Mike, you can't make a silk purse out of a sow's ear. These kids are culturally deprived. I feel more like a caretaker than a [teacher]" (p. 60).

25 Greig (2011) summarizes Paul Willis' 1981 ethnography of working-class teenage boys in the UK, which is reminiscent of Bushler Bay, but also uses class as a descriptor:

> The study highlighted the young men's active and self-conscious cultivation of a 'traditional' working-class masculinity as a form of resistance to being labelled failures in the context of the middle-class aspirational values of school. Their gender practice became a source of class dignity that, in its rejection of education, only served to reproduce capitalist relations by ensuring working-class kids stayed in working-class jobs.
>
> *(p. 227)*

26 McKay and Lucero-Prisno (2012) refer to a 2006 study in the US:

> Working class labourers often celebrate a masculine 'pigness'—coarse, physically tough and aggressively heterosexual—to set themselves apart from more refined, effete professionals who lack 'real' manhood.
>
> *(p. 21)*

27 Besides the school, Public Employees (mostly men) worked in the US Forest Service, Washington State Shellfish Lab, Washington State Fish Hatchery and the US Park Service. Women tended to conform to the values and behaviours aligned with their husbands' work, regardless of their own upbringing.
28 Reading the auto-ethnographic account of a Finnish *woman*'s attempts to control her tears when confronted with devastating personal news (Katila 2019) can help us imagine the difficulty many men may have doing so. Kenyah men are allowed to cry (e.g., Chapter 6).
29 Cf. Silberschmidt's (2011) work in East Africa, showing how men who are unable to perform their socially defined roles as breadwinners turn to violence and extra-marital sexuality to maintain their self-respect as men. See also Alcaraz and Suárez (2006) on this pattern in Colombia or Barker and Ricardo (2006) in sub-Saharan Africa generally; also noted in the literature survey by Morris and Ratajczak (2019).
30 The same symbol, an apron, with a similar implication of loss of masculinity, is reported for aged men in urban Mexico (Varley and Blasco 2001).
31 Compare Hendriks' (2014) observations of Danish, French and Spanish loggers in a logging camp in the Democratic Republic of Congo:

> The STIHL [a chainsaw manufacturer] pictures themselves play with stereotyped images of (white) masculinity and femininity and reconfirm the image of the logger as the paradigmatic masculine male and the chainsaw as the ultimate phallic object.
> *(p. 220)*

32 DeMille and Lyons (2016) report, based on a study in Coastal British Columbia, that,

> Historically manual tree falling has had a disproportionately high rate of fatal injuries. Syngatur (1998) reported that the rate of fatal injuries per 100,000 workers for manual tree fellers was 128.7 as compared to aircraft pilots (83.3), truck drivers (27.9), and farm occupations (27.5).
> *(p. 433)*

33 The emphasis on courage can be even stronger in areas where forest fires are common.
34 Like the Russian youth described by Walker (2016):

> With apparently secure forms of low-skilled manual employment awaiting them after school, the lads [in a 1977 ethnography] were able confidently to reject the 'educational exchange' (of subservience for qualifications), and valorized manual over mental labor.
> *(p. 51)*

35 Apparently similar to McLeod (2016): "The stereotype of the 'Kiwi' male as pioneering, self-sufficient and strongly tied to the land is still a potent motif found in advertising and popular culture" in New Zealand (p. 227).
36 Connell links 'marginalized masculinities' with class or race. But within the Local context, bureaucrats, more highly valued in the broader society, are marginalized as unable or unwilling to 'perform' proper manhood.
37 'Peter' is a colloquial American term for the penis.
38 A pun on this common brand of log truck.
39 Robinson and Hockey (2011) note a similar masculine fascination with big trucks: firemen in the UK joined the service "'*to ride the wagons*' (fire engines)" [italics in original, p. 102]. The authors also describe hypersexuality similar to what Local women attribute to loggers.
40 Barker et al. (2011) report men's awareness of this attraction:

> In one telling example, young men in a group session in Brazil said that if they became more sensitive, or gender-equitable, they wondered if they would convince young women in the community to go out with them. Young men

> argued, and young women in the group confirmed, that young women often like to go out with the 'bad guys'.
>
> *(p. 179)*

This 'naughty boy' syndrome is also found in Indonesia (see Chapters 4 and 5) and elsewhere.

41 Compare Miller (2009) on the Rarámuri of northern Mexico:

> Rarámuri men gain respect when they take part in birth, and it is not shameful to talk about it because public expression of the fulfillment of a social obligation is part of the moral life of all Rarámuri. In this manner, men's active role in the reproductive process reaffirms . . . the egalitarian and interdependent nature of the marriage bond as an essential feature of Rarámuri social organization.
>
> *(p. 341)*

References

Alcaraz, Fredy Hernán Gómez, and Carlos Iván García Suárez. 2006. "Masculinity and Violence in Colombia: Deconstructing the Conventional Way of Becoming a Man." In *The Other Half of Gender: Men's Issues in Development*, edited by Ian Bannon and Maria C. Correia, 93–110. Washington, DC: The World Bank.

Altork, Kate. 1995. "Walking the Fire Line: The Erotic Dimension of the Fieldwork Experience." In *Taboo: Sex, Identity, and Erotic Subjectivity in Anthropological Fieldwork*, edited by Don Kulick and Margaret Willson, 107–139. London and New York: Routledge.

Anderson, Eric. 2009. *Inclusive Masculinity*. New York: Routledge.

Ballard, Heidi L., and Lynn Huntsinger. 2006. "Salal Harvester Local Ecological Knowledge, Harvest Practices and Understory Management on the Olympic Peninsula, Washington." *Human Ecology* 34:529–547. https://doi.org/10.1007/s10745-006-9048-7.

Barker, Gary, Marcos Nascimento, Christine Ricardo, Marianna Olinger, and Márcio Segundo. 2011. "Masculinnities, Social Exclusion and Prospects for Change: Reflections from Promundo's Work in Rio de Janeiro, Brazil." In *Men and Development: Politicizing Masculinities*, edited by Andrea Cornwall, Jerker Edström, and Alan Greig, 170–184. London: Zed Books.

Barker, Gary, and Christine Ricardo. 2006. "Young Men and the Construction of Masculinity in Sub-Saharan Africa: Implications for HIV/AIDS, Conflict and Violence." In *The Other Half of Gender: Men's Issues in Development*, edited by Ian Bannon and Maria C. Correira, 159–193. Washington, DC: The World Bank.

Blanchard, Kendall. 1974. "Basketball and the Culture Change Process: The Rimrock Navajo Case." *Council on Anthropology and Education* 5:8–13.

Brandth, Berit, and Marit S. Haugen. 2000. "From Lumberjack to Business Manager: Masculinity in the Norwegian Forestry Press." *Journal of Rural Studies* 16:343–355.

Burwell, Lakota. n.d. "Environmental Factors Influencing Fire Behavior in the Olympic Mountains." *Humboldt State University Capstone Project 476*. www.nps.gov/olym/learn/management/fire-history.htm.

Carroll, Matthew Stephen. 1985. *Community and the Northwestern Logger: Continuities and Change in the Era of the Spotted Owl*. Boulder, CO: Westview Press.

Colfer, Carol J. Pierce. 1975. "Bureaucrats, Budgets, and the BIA: Segmentary Opposition in a Residential School." *Human Organization* 34 (2):149–156.

Colfer, Carol J. Pierce. 1977. *Women's Communication and Family Planning in Rural America: The Case of Bushler Bay*. Edited by D. Lawrence Kincaid. Vol. 4, *Case Studies in Family Planning*. Honolulu, HI: EWCI (Case Study #4; reprinted 1978).

Colfer, Carol J. Pierce. 1978. *That All Men Are Created Equal . . .* Cambridge, MA: Abt Associates.

Colfer, Carol J. Pierce. 1983. "On Communication Among 'Unequals'." *International Journal of Intercultural Communication* 7:263–283.

Colfer, Carol J. Pierce. 2018. "Before and After the 'Timber Wars': Context, Change and Potential Collaboration on the Olympic Peninsula." *Humboldt Journal of Social Relations* 40:137–162.

Colfer, Carol J. Pierce, Lee Cerveny, and Susan Stevens Hummel. 2019 "Using Rapid Rural Appraisal Tools to Explore Gender and Forests in the Global North." *Human Organization* 78 (1):12–27.

Colfer, Carol J. Pierce, with A. Michael Colfer. 1978. "Inside Bushler Bay: Lifeways in Counterpoint." *Rural Sociology* 42 (2):204–220.

Colfer, A. Michael, and Carol J. Pierce Colfer. 1979. *Big Money, Small Change: Cultural Paradigms in an American Village*. Cambridge, MA: Abt Associates.

Cornwall, Andrea. 2016. "Introduction: Masculinities under Neoliberalism." In *Masculinities Under Neoliberalism*, edited by Andrea Cornwall, Frank G. Karioris, and Nancy Lindisfarne, 1–28. London: Zed Books.

DeMille, Gregory John, and Kevin Lyons. 2016. "Unexpected Events When Manually Falling Trees in Coastal British Columbia." *Forestry Science* 62 (4):433–439. https://doi.org/10.5849/forsci.15-164.

Greig, Alan. 2011. "Anxious States and Directions for Masculinities Work with Men." In *Men and Development: Politicizing Masculinities*, edited by Andrea Cornwall, Jerker Edström, and Alan Greig, 219–235. London: Zed Books.

Headwaters Economics. 2012. *The Economy of the Olympic Peninsula and Potential Impacts of the Draft Congressional Watershed Conservation Proposal*. Bozeman, MT: Wild Olympics Coalition.

Hendriks, Thomas. 2014. "Race and Desire in the Porno-tropics: Ethnographic Perspectives from the Post-colony." *Sexualities* 17 (1/2):213–229.

Katila, Saija. 2019. "The Mothers in Me." *Management Learning* 50 (1):129–140. https://doi.org/10.1177/1350507618780653.

McKay, Steven, and Don Eliseo Lucero-Prisno III. 2012. "Masculinities Afloat: Filipino Seafarers and the Situational Performance of Manhood." In *Men and Masculinities in Southeast Asia*, edited by Michele Ford and Lenore Lyons, 20–37. New York: Routledge.

McLeod, Carmen. 2016. "Ducks, Dogs, and Men: 'Natural' Masculinities in New Zealand Duck Hunting." In *Masculinities Under Neoliberalism*, edited by Andrea Cornwall, Frank G. Karioris, and Nancy Lindisfarne, 227–243. London: Zed Books.

Meyer, John C. 2000. "Humor as a Double-Edged Sword: Four Functions of Humor in Communication." *Communication Theory* 10 (3):1–33.

Miller, Janelli F. 2009. "Husband-Assisted Birth Among the Rarámuri of Northern Mexico." In *Reconceiving the Second Sex: Men, Masculinity and Reproduction*, edited by M. C. Inhorn, T. Tjornhoj-Thomsen, H. Goldbert, and M. L. C. Mosegaard, 327–348. New York: Berghahn Books.

Morris, Edward W., and Kathleen Ratajczak. 2019. "Critical Masculinity Studies and Research on Violence Against Women: An Assessment of Past Scholarship and Future Directions." *Violence Against Women* 25 (16). https://doi.org/10.1177/1077801219875827.

Newsweek. 1974. "The Riskiest Job of All." *Newsweek*:64.

Pascoe, C. J. 2007. *Dude, You're a Fag: Masculinity and Sexuality in High School*. Berkeley, CA: University of California Press.

Pingree, Melissa R. A., and Thomas H. DeLuca. 2018. "The Influence of Fire History on Soil Nutrients and Vegetation Cover in Mixed Severity Fire Regime Forests of the Eastern Olympic Peninsula, Washington, USA." *Forest Ecology and Management* 422:95–107.

Reed, Maureen G. 2003. "Marginality and Gender at Work in Forestry Communities of British Columbia, Canada." *Journal of Rural Studies* 19:373–389.

Robinson, Victoria, and Jenny Hockey. 2011. *Masculinities in Transition*. New York: Palgrave Macmillan.

Silberschmidt, Margrethe. 2011. "What Would Make Men Interested in Gender Equality? Reflections from East Africa." In *Men and Development: Politicizing Masculinities*, edited by Andrea Cornwall, Jerker Edström, and Alan Greig, 98–110. London: Zed Books.

Syngatur, E. F. 1998. "Logging Is Perilous Work." *Compensation and Working Conditions* (winter):9. www.bls.gov/opub/mlr/cwc/logging-is-perilous-work.pdf, accessed 12 February 2016.

Tantam, William. 2016. "Football Field, Bar and Street Corner: Sports, Space and Masculinities in Rural Jamaica." In *Masculinities Under Neoliberalism*, edited by Andrea Cornwall, Frank G. Karioris, and Nancy Lindisfarne, 213–226. London: Zed Books.

Varley, Ann, and Maribel Blasco. 2001. "Exiled to the Home: Masculinity and Ageing in Urban Mexico." In *Men at Work: Labour, Masculinities, Development*, edited by Cecile Jackson, 115–138. Portland, OR: Frank Cass with EADI (European Association of Development Research and Training Institutes, Bonn).

Walker, Charlie. 2016. "In Search of 'Stability:' Working-class Men, Masculinity and Wellbeing in Contemporary Russia." In *Masculinities Under Neoliberalism*, edited by Andrea Cornwall, Frank G. Karioris, and Nancy Lindisfarne, 51–65. London: Zed Books Ltd.

Westberg, Lotten, and Stina Powell. 2015. "Participate for Women's Sake? A Gender Analysis of a Swedish Collaborative Environmental Management Project." *Society & Natural Resources* 28 (11):1233–1248. https://doi.org/10.1080/08941920.2015.1014594.

4

MASCULINITIES AND MUTED GENDER IN BALI AND KALIMANTAN

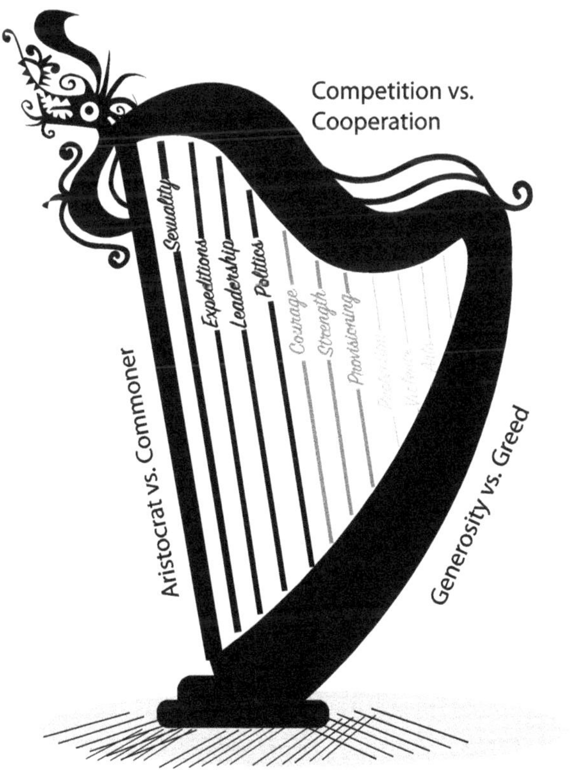

A Kenyah
Harp Example

Introduction and reflections

I arrived in Indonesia for the first time in June 1978, 33 years old, alone and without friends or language capability. I did have the names of some people that friends in Hawaii thought would help me. On arrival, in the middle of the night, I remember my excitement tinged with a bit of anxiety as the taxi driver and I struggled to communicate and find my friend's house (no Google Maps, of course) in the sprawling Jakarta metropolis. I knew only that it was 'near the prison'.

My first surprise was the care these people, including men, took of me. Cars drive on the left side of the street in Indonesia, and I had difficulty at first remembering to look the proper direction as I moved out into the chaos of Jakarta's traffic. One young man I'd just met at Jakarta's Population and Family Planning Bureau (BKKBN, *Badan Kependudukan dan Keluarga Berencana*) held my hand as we crossed streets, recognizing my incompetence and saving me from certain death. Another stranger took me in hand on my trip to Yogyakarta, helping me find a place to stay and the address of a mask maker who would lead me to the young Nancy Peluso, already long resident in Indonesia. I was intrigued by the kindness I was shown by men who didn't know me, and without obvious ulterior motives.

My first visit of any length was in Bali, an interlude in which I learned rudimentary Indonesian and began to acclimate to the country—in its late 1970s state. The lack of concerns about privacy, people's willingness to share their lives, their ideas, their way of life, all were a delight for an anthropologist tired from the suspicion and prickliness of Bushler Bay's residents. Bali, as part of 'inner' Indonesia, in Geertz (1963) terms, was tightly integrated with national and even international policy, densely populated, nationally recognized as 'civilized', hierarchical and Hindu—all unlike the Kenyah situation discussed later in this chapter. I would encounter some of these qualities again, in Sitiung (Chapter 5).

The Bali of 1979 was characterized by different axes of power than today's Indonesia. The government was super-centralized, with power clearly in the hands of President Soeharto and his cronies. Java and Bali were differentiated very clearly from the 'Outer Islands' (Chapter 5). There was little or no awareness, to my knowledge, of gender as an issue, but family planning was one of the central government's primary concerns. And many governmental policies included sexist assumptions about gender (discussed by Elmhirst 2011, for the transmigration programme, and in Chapter 5)—men as heads of households, women as homemakers, men as land owners, breadwinners, politicians, etc. The intersection of gender with other social categorization is obvious, with caste, educational level, race, nationality and occupation all relevant in the discussion in this chapter (indeed, throughout this book).

I introduce this urban, only peripherally forested place, for three reasons. The first is to provide continuity in my unfolding understanding of masculinities . . . as I moved from the very polarized gender world of the US to Indonesia, where many groups did not so clearly differentiate.[1] The second is to take advantage of the opportunity to briefly highlight, yet again, the divergences in masculinities, even within one country. And the third is to introduce briefly international masculinities—through donor actions, competition among scientists and institutions and different scientific preferences—more fully explored in Chapter 6. The willingness to accept the views of donors and outside experts, and act on their suggestions, was much higher in the 1970s than it is today. Again, I set the stage by explaining a bit about the biophysical environment.

Bali: my introduction to Indonesia[2]

In Bali I lived an urban life, working among elites. The area around Denpasar, a plain, had more than 500 people per square kilometre, yet still the island was considered a tourist's paradise.[3] Rainfall in the city was between 1500 and 2000 mm/year, higher in other areas of Bali. Not devoid of tropical rainforest, the nearest was north of the city: moist deciduous forest (the most common forest type on Bali), with some areas of evergreen and semi-evergreen rainforest further north, most above 500 m elevation. When Whitten, Soeriaatmadja, and Afiff's (1996) book was published, only 18% of Bali was forested (a bit over 100,000 ha), though there was at that time a probably unrealistic plan to expand that to 31%. *Manilkara kauki* was the most important indigenous species found in lowland areas (short, stocky and used for woodcarving); 5000 ha of this species along with teak and other timber species were planted by the Indonesian government in the early 1960s. Far more ubiquitous in the landscape were the beautiful rice terraces that lined hillsides.

Entranced by the beauty of the island, I was also immediately struck by the deep dark pools of men's (and women's) eyes, which seemed—in my youthful naiveté—to allow me to see directly into their souls. There was an openness, a kind of social trust and caring, that were new to me and in which I happily basked for the two months I stayed there. Social interaction felt like being bathed in a pool of warm water; people were gentle and helpful, with no hint of the suspicion I'd lived with for my three years in Bushler Bay or my nearly seven years in Turkey. It was a soothing place.[4] But my understandings of masculinity there are far more superficial than in the places where I stayed longer.

I became very close to my hostess, a wealthy and beautiful woman about my age. Her husband, a physician, had been in graduate school with me and had graciously invited me to stay in his 'empty' house with his wife.[5] I was also connected with the highest level officials, almost all men, within Bali's very successful family planning programme; with consultants who came through, many from Australia; with anthropologists briefly in town from the field; and with officials and professors at Udayana University. I quickly encountered the relevance, and different manifestations, of several familiar harp strings.

Sexuality was a key concern. My hostess had considerable interest in her own, her friends' and her family's (and my own) sexuality. Her husband, away in Hawaii, had been so concerned to ensure her fidelity that he'd installed his mother in his office, instructed three of his female relatives to sleep outside her bedroom door every night, hired his friends to work in the hotel where she worked and removed her IUD—all as preventive measures.

Although extra- and pre-marital sex were disapproved and the subject of much discussion, both were considered common. Men flirted with me incessantly, though never harassed. My hostess had many questions about American sexuality, and the Balinese loved discussing extra-marital affairs, some with serious consequences for the participants' careers. In one then-recent case, two married doctors had an affair, which became widely known. They both left Bali, going to different cities on Java, both their licences to practise in Bali revoked for two years. When I asked why that had been necessary, an anthropologist experienced in Bali told me, "The Balinese say they [the doctors] couldn't control themselves, and that you can't trust doctors who can't control themselves" (notes, 4 August 1979). In both these examples, sexuality and control were closely linked. My hostess told me,

> Balinese believe that men are at their strongest (*kuat*) for sex when they are about 40 and that they begin to look at younger women when their wives are about 40. Women are believed to go downhill sexually after that.
>
> *(notes, 20 July 1979)*

The most dramatic surprise related to sexuality and reproduction came toward the end of my stay, by which time I'd learned a little Indonesian. I was invited by the Dean of Udayana University's Medical School, with which I was loosely affiliated, to attend a vasectomy clinic in Karangasem to the east (notes, 11 August 1979). Some distance from Denpasar, we came to a small, simple clinic where 42 men were seated in rows on benches on a veranda. They looked friendly, comfortable and at ease. I was surprised to learn that they were waiting to have vasectomies. I kept imagining the likely reaction should a man from Bushler Bay be invited to have his vasectomy so publicly. But more surprises were to come. Inside the building, there were five operating tables, on each of which lay a man with a green cloth covering all of his body, save his head and his genitals. My small group of medical professionals, all men, engaged the patients in friendly conversation. The latter responded with apparent ease and no hint of embarrassment, smiling with their open friendly eyes at me as well. The next surprise was the appearance of a television crew, which proceeded to interview the men and film their vasectomy operations. These images were shown on television that evening for all to see. There was obviously a very different attitude about vasectomies (and masculinity) in Bali than I'd encountered in Bushler Bay or with my own grandfather. Men's enthusiasm to undergo this procedure may have been linked to its perceived modernity, to 'things western' and national priorities at that time.[6]

Family planning of all kinds was organized very publicly in Bali. There was a strong push from Indonesia's President Soeharto encouraging family planning throughout the country—so strong that some international actors criticized the government for forcing people to practise birth control.[7] The vasectomy clinic was attended by high-level government officials from the district, the military, the police, the Health Ministry and BKKBN. The BKKBN director had a direct mandate from the president, and thus was able, unlike most government leaders, to involve other ministries in his efforts.

The Balinese had cleverly incorporated family planning into their *Banjar* system. Each neighbourhood had a *Banjar* composed of local men, who met regularly to organize and monitor community affairs. Unlike in Bushler Bay, politics was primarily in the hands of men. To this task was added the responsibility of ensuring compliance with the family planning programme. Spray painted on the front of each house was the birth control method that family used. I detected no hint that any kind of family planning was private or embarrassing in any way; nor did it seem associated with masculinity (or the lack thereof). Discussion and approval of family planning were very common among both men and women, though its distribution was in the hands of men.

As with Bushler Bay's Public Employees, hierarchy was important in Bali, but its manifestation differed greatly. The hierarchies of the various agencies and departments were further complicated by the existence of a separate and explicit caste system; men were expected to marry within their own or a lower caste. Dr. Ngurah Bagus, an Udayana University anthropologist, noted that the

> lower castes have been fighting the caste system a lot. There used to be restrictions on marriage, speech and eating together. Now only on ceremonial occasions is there separate eating. And marriage is becoming more common between castes, though families are still not happy. It particularly messes things up on ceremonial occasions, if a woman marries beneath herself.
>
> *(notes, 13 August 1979)*

My hostess had done just that and had had to wait eight years before she and her husband were able to marry, perhaps because of the caste difference.

Language, as Drs. Bagus notes, also remained a problem, as there are refined (*halus*) and coarse (*kasar*) forms of the Balinese language:

> In order to avoid the confusion of when to use *halus* and *kasar* Balinese, lower caste people use Bahasa Indonesia. He said it's no problem for the upper castes, only for highly placed lower castes. He said *sudra*, the word I'd learned for lower castes, is an insult; we must use *jaba* or *orang kebanyakan* ['most people'].
>
> *(notes, 13 August 1979)*

These national hierarchies were compounded among these elites by relations between Indonesia and potential donors—a topic I explore in some detail, in anticipation of Chapter 6. In 1979, Indonesia had only recently emerged from three centuries of Dutch colonial rule. Many of those in power retained a sense of the superiority of all things western and white, and formal education. Although the funds available from external donors were very much desired and welcomed on their own, there remained also a sense of subservience and extreme respect granted to people from the US and western Europe.[8]

Having come from the University of Hawaii to complete my master's in public health, I was affiliated with an impressive and well-respected American professor, active both within our university and within USAID, whom I'll call Dr. Q. His expectation of control in his relations with the Balinese was clear:

> Dr. Q . . . is squiring around some hotshot from AID. I asked about the nutrition mess, and he said it was all ok, it had just stopped progressing while he was in the US. He said at one point he got so mad at Pak A [a BKKBN official] he told him he was going to take back his diploma. [Another BKKBN official] told Dr. Q that the workshop will have to be postponed because AID hasn't paid the money yet. But Dr. Q's gonna bring [an American doctor] as planned and pretend he was supposed to be in on the planning.
>
> *(notes, 25 July 1979)*

At the same time, Dr. Q showed his commitment to a caring or protective role, both with the Balinese and with me. In one case, he seemed to opt for my benefit over that of his hosts: he insisted that I be included in a planned budget as a foreign, rather than a local, consultant. I would thereby get much more money than the Balinese counterpart consultants. But his reasoning showed his real concern for project success: if I weren't available to do it (a distinct possibility), that larger sum could be shifted to [more] Balinese consultants.

There was also evidence among the Balinese BKKBN bureaucratic men of a strong preference for mathematical and experimental approaches to research (as in the US). B [a junior BKKBN official] and I

> met to go over the methodology section I'd written. He was not at all shy to explain what he thought should be in the proposal. He prefers mathematics to reading or memorizing. Teaches nutrition and statistics.
>
> *(notes, 9 August 1979)*

Drs. C [anthropologist] says he himself

> always attacks the people at *Kedoktoran* [Ministry of Health] and Public Health because they don't make use of the knowledge of other disciplines. He said, with sex education, we [at Udayana University] have an Education Department, but they don't use it. They're too narrow minded. They think they're better than us. He was as vehement as I've seen Balinese get. It was a sentiment I recognized well [in interactions between qualitative and quantitative researchers].
>
> *(notes, 13 August 1979)*

Hierarchy strikes me as a stronger social structural principle in Bali than gender differentiation.[9] From an international perspective, I was, for instance, more fully and readily included in the work of BKKBN than an equally qualified Australian man,[10] who encountered one roadblock after another. I thought my access derived from my affiliation with Dr. Q, who had funds and knowledge he was trying his best to share. The fascination many Indonesians had with white women could also have played a role. The neglected Australian (L) recounted his experience to me:

> He . . . met Dr. Q when he'd been in Indonesia for three days. Was introduced by someone who was urging Dr. Q to use this demographer's expertise. L said he felt like he was caught in a power play, and that Dr. Q had felt L was messing around in his backyard. Sounded like a quite unpleasant interchange, and he said, "neither of us has gone out of our way to mend fences". He thought there was a rivalry between the Hawaii mafia and the ANU crowd of demographers.
>
> *(notes, 14 August 1979)*

The high-status men with whom I was primarily working took their leadership—another common masculine harp string—responsibilities seriously. The head of BKKBN in Bali spoke at length about his leadership philosophy and experience. He

> expressed a philosophy of working behind the scenes, getting people's cooperation, stepping out of the limelight (e.g., writing the speech the governor gives, arranging the [vasectomy] camp that's officially the Bupati's [district leader's] shindig). He believes in delegating so that the thing is *theirs*, not his. He also believes in making use of the powers that be.
>
> His organization is outside the regular bureaucratic structure, and he is therefore, with the active help of the [central] Government, able to coordinate the Ministry of Education, the Ministry of Information and the Ministry of Health. He attributes a lot of his success to his utilization of the powers that be to back his programs. He also recognizes the importance of making the program belong to the people it serves. Both these concepts came through again and again. He said he speaks *halus* to villagers. He also spoke Balinese to them, to be closer to them, he said. Both he and the Bupati asked the villagers if they had been pressured or if this [having a vasectomy] was of their own free will (for my benefit? Or to show each other?). I noticed that the Bupati gently helped a man off the operating table in a caring way. He also was the first person to offer blood. Both reiterated the necessity for leaders to be the pioneers and were happy when a *kelian* [*banjar* leader] had been the first vasectomy acceptor, though neither of them had had a vasectomy. The Bupati said

> his wife was one of the first tubectomies, and [the BKKBN head] said his wife didn't want him to.
>
> *(notes, 11 August 1979)*

To sum up, some of the key concepts observed among men in Bali included sexuality, control, leadership, competitiveness and comfort with hierarchy—this latter more complex and pervasive than that seen among Public Employees in Bushler Bay. Among the internationally involved, again control and competitiveness were evident, as was a preference for math and science, the quantitative and experimental over qualitative observational research methods. There was also a comfort among these men with issues of sexuality, at least as pertained to family planning, that would not recur among my colleagues in the natural resource fields discussed in Chapter 6.

All in all, this detour in Bali began my process of moving out of an American polarized view of gender to a much more flexible one than I had ever encountered before. It introduced me to Indonesian science, governmental bureaucracy and donor relations, and primed me for the new and unearned respect I would be given because of my colour, comparative wealth[11] and academic degrees. These latter overshadowed my gender in these contexts.

Masculinity among the Uma' Jalan Kenyah Dayaks[12]

Bali had been a brief introduction to one partial version of masculinity in Indonesia. Long-term research took me to East Kalimantan in September 1979. I have no words to convey my utter delight at the prospect of going deep into a tropical forested world, a world totally new to me. I begin this journey with a description of the biophysical world of the Kenyah in the 1970s and '80s.

Introduction to Long Segar, East Kalimantan

Long Segar is in the middle of East Kalimantan, situated along the Telen River (see Figure 4.1). An unusually forest-rich environment at that time,[13] it was surrounded most closely by a patchwork of upland rice fields, early and late secondary forest. As the distance from the village increased, so did the amount of old growth forest. Population density in the area was estimated at 2/km^2 (Directorate General of Forest Utilization 1989). Dipterocarps, the dominant commercial species, could reach 45–60 m or occasionally taller and sometimes comprised as much as 10% of all trees and 80% of all emergents.

> The combination of very high stocking of trees with huge boles, commonly 20m long or more, and of relatively light weight, has encouraged extensive exploitation of dipterocarp forests throughout Southeast Asia.
>
> *(MacKinnon et al. 1996, p. 177)*

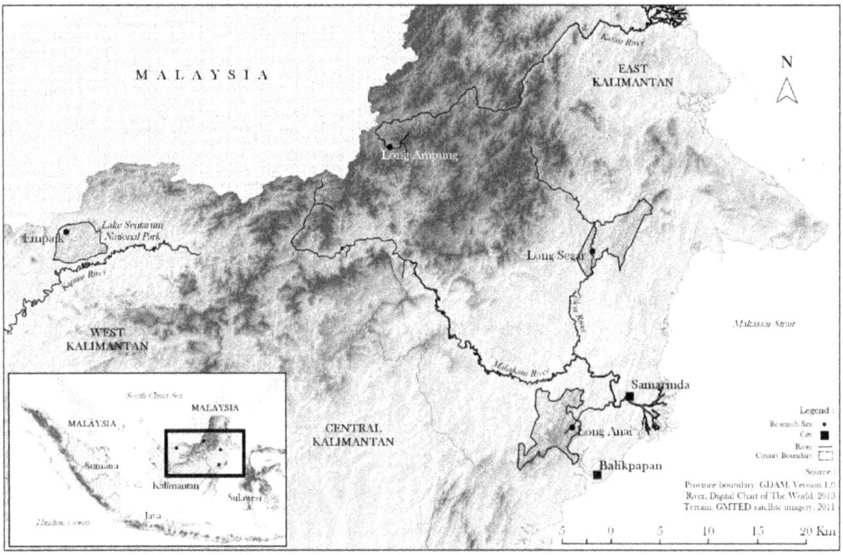

FIGURE 4.1 Map of East Kalimantan research sites.

The Long Segar area was a perfect example. Timber companies were operating all around the community, which had been declared part of Indonesia's Forest Estate. The biggest concession (in which Long Segar itself was located) was managed by the American Georgia Pacific Company (turned over to PT Kiani Lestari after the fires of 1982–1983).

Although the very centralized Ministry of Forestry officially managed the area from Jakarta, day-to-day management was in the hands of the American company, with only an occasional visit from forestry officials. This concession was categorized as production forest and was logged primarily for its valuable dipterocarps (~ten trees/ha). Although the US Foreign Corrupt Practices Act had already been passed (1977), there was a widespread belief that all the timber companies in the area[14] bribed officials and followed governmental regulations only along good roads that the companies maintained close to base camps. In these easily accessible forest areas, prescribed management (e.g., cutting guidelines, annual allowable cut limits, replanting, etc.) was followed. Visiting officials were not usually anxious to take long, hot, bumpy rides over muddy or dusty logging roads into more distant areas.[15] The companies were able to do as they chose—what was most profitable—in the remainder of the concessions. They were also able, in effect, to bribe local leaders to accept whatever management practices the companies chose, framed as fulfilling their legal obligations to contribute to the well-being of local communities (see Chapter 7, for similar narratives among oil palm companies).

The main dipterocarps included *Dipterocarpus, Dryobalanops* and *Shorea* (emergents) and Hopea and Vatica, smaller trees. The legumes included *Dialium, Koompassia* and *Sindora* (all emergents). One particularly interesting species was the magnificent *tanyit* (*Koompassia excelsa*), with its white bark and high, spreading canopy (one example in nearby Sarawak reached 83.82 m)—the tallest broadleaf rainforest tree in the world and home to honeybees.

Another key species in the area was *beli'en* (*Eusideroxylon zwageri*), or ironwood, used in much construction locally (see Colfer, Peluso, and Chin 1997, for lists of various forest products—timber, fibres, foods and medicines—used by the Kenyah). MacKinnon and MacKinnon (1986, quoted in MacKinnon et al. 1996) note that by 1986, 60% of the original area of ironwood forest had been lost, as had a third of lowland dipterocarp forests in Kalimantan generally (p. 398). Another important species used by the Kenyah was rattan (especially *Daimonorops spp., Calamus javensis, C. caesius, C. trachycoleus, Korthalsia echinometra, K. rigida* and *Plectocomiopsis geminiflora*), though used more commercially by the Kutai community of neighbouring Kernyanyan (Colfer, Peluso, and Chin 1997).

In 1981, log production in East Kalimantan was 2,856,560m^3.

> In 1978, this province with less than 1% of Indonesia's population, produced nearly 25% of the country's total export earnings, mainly from timber and petroleum products.
>
> *(McKinnon et al. 1996, p. 401)*

This forest, in 1979 well protected by the humidity and extent of old growth forest, began to suffer from devastating forest fires whenever El Niños of any severity occurred (e.g., 1982–1983, 1997). Although there was a serious El Niño in 1972–1973, which caused great difficulties for the Kenyah, the main problem then was a failed rice crop, rather than a totally burned and devastated landscape (Colfer and Dudley 1993).

I initially went to Long Segar, accompanied by a senior forestry student from Mulawarman University (Albar Azier), prepared to conduct a conventional ethnographic study. I immersed myself in local life, living with the village headman's family of 14, from October 1979 to August 1980. In February, I was joined by my ten-year-old daughter and soon-to-be-ex-husband (the latter staying until May). I spent May 1980 in Long Ampung, the village from which Long Segar residents had moved between 1963 and 1972.

Soon after my arrival, I was struck by the degree to which the differentiation between men and women was muted,[16] vis-à-vis the situation in the rural US. I had just come from a community where men and women were seen as polar opposites (as Newton 1977 found in Hawaii); in Long Segar, they were seen as quite similar. We might say that the Long Segar harp of masculinity is smaller than that of Bushler Bay, or perhaps its sounds have been dampened to softer tones. Or perhaps the harp is simply one of humanity rather than of masculinity per se, with men plucking somewhat different strings and clusters, creating somewhat different songs. Mashman and Nayoi (2015) reflect another common academic view of Indonesian gender relations (at least outside Java and Bali): "Complementarity seems to be the more relevant mode of analysis, as opposed to asymmetry and inequality" (p. 960).[17]

In 1980, I conducted a cognitive mapping study (called 'Galileo'), designed to quantify people's perceptions of men, women and natural resource concepts.[18] The resulting 'map' and matrices showed the average cognitive 'distances' people attributed to pairs of concepts in a multidimensional space. With ten points being the 'length' of the cognitive measuring tape, respondents assessed *men* and *women*, on average, to be two points apart—providing some quantification of the widely recognized common humanity of Kenyah women and men. Another interesting concept was *good*. The *women–good* link was assessed, on average, as 1.8 (along with *rice field–good*; only *work*, at 1.4 had a closer link to *good*); the *men–good* link was 2.2. The greatest distance mentioned in this adult dataset was between *trade* and *child*, 7.4; the smallest pairs were *woman–child* and *cottage industry*[19]*–work* (see Figure 4.2 and Table 4.1). .

Uma' Jalan Kenyah had lived for about a century in Long Ampung, in the remote interior of Borneo, bordering Malaysia (see Figure 4.1). In 1963, a small Protestant segment of the community had moved to the somewhat more accessible

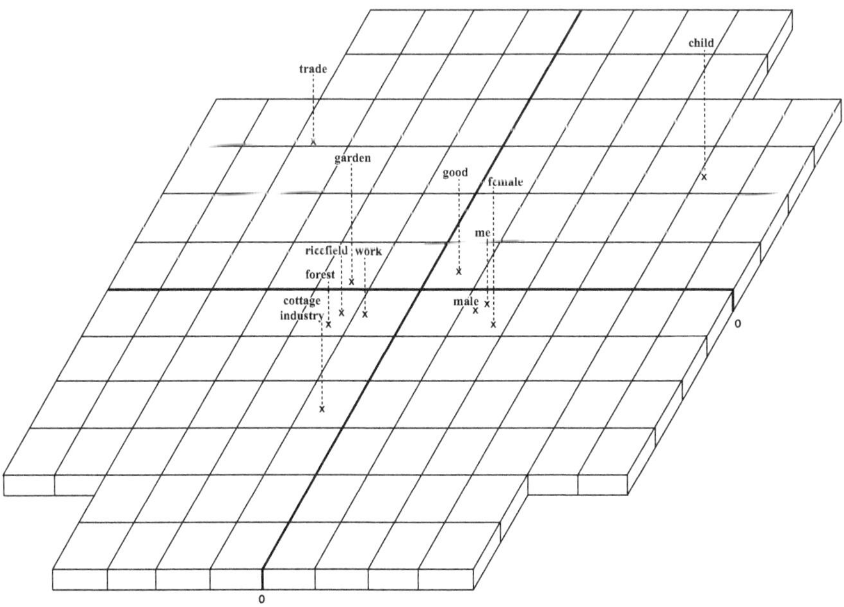

FIGURE 4.2 Map showing cognitive distances between men and women in Long Segar, East Kalimantan, 1980.

Source: C. J. P. Colfer, "Women of the Forest: An Indonesian Example." In *Women in Natural Resources: An International Perspective*, edited by Stock, Force and Ehrenreich. Moscow, Idaho, 1982.

TABLE 4.1 Galileo means matrix—cognitive mapping study, Kenyah adults, Long Segar, 1980

	Forest	Rice field	Garden	Male	Female	Child	Work	Me	Good	Trade	Cottage Industry
Forest	0.0										
Rice field	1.5	0.0									
Garden	1.7	2.2	0.0								
Male	3.0	2.6	4.0	0.0							
Female	3.7	2.0	1.7	2.0	0.0						
Child	6.4	6.1	6.0	3.2	1.2	0.0					
Work	2.4	1.4	1.4	1.5	1.4	5.7	0.0				
Me	3.0	2.1	4.0	1.9	1.8	2.0	1.4	0.0			
Good	2.4	1.8	2.2	2.2	1.8	2.4	1.4	3.1	0.0		
Trade	5.1	4.3	4.2	6.3	6.8	7.4	4.3	6.4	5.1	0.0	
Cottage Industry	3.1	2.6	3.7	3.2	1.4	7.0	1.2	2.7	1.9	6.4	0.0

Source: C. J. P. Colfer, "Women of the Forest: An Indonesian Example." In *Women in Natural Resources: An International Perspective*, edited by Stock, Force and Ehrenreich. Moscow, Idaho, 1982

Long Segar. The government had in 1975 declared Long Segar a Resettlement Village, providing the people with various 'inputs' (seeds, farming tools, extension services and pressure to plant paddy rice), like the transmigrants discussed in Chapter 5. This change also meant that the community's presence there was acknowledged by the government; formal certificates of land ownership were promised, though not forthcoming. The core of Kenyah subsistence was swidden agroforestry, supplemented by hunting, fishing, gathering and periodic male wage labour, sometimes with the US-based Georgia Pacific timber concession two hours downriver.[20]

Although men and women participated almost equally in swiddening labour—the most common forest-dependent agricultural method in the tropics—in Long Segar (Colfer 1981),[21] rice production was considered women's profession. Men emphasized other elements of the subsistence system (hunting, fishing, gathering and periodic wage labour). As with many other groups, women were more involved in reproductive activity (child- and eldercare, cooking, cleaning) than men. But the division of labour was flexible. Men could also often be seen washing clothes, cooking and taking care of children. Although men usually piloted canoes, women also took the helm. There was no taboo against women doing what men more typically did or vice versa.

> Kenyah seem to be singularly nonjudgmental about either sex performing tasks that are most commonly undertaken by the other. The harsh . . . ribbing that accrues in some places to people who deign to engage in activities reserved by that group for the opposite sex is entirely absent. The Kenyah are aware of sex role rigidity among some other ethnic groups they have

> encountered. Kayan men, for instance, were described to me as responding to floor sweeping in the same way that American men have learned to respond to dishwashing and diaper changing. But the Kenyah merely chuckle about such cultural foibles.
>
> *(Colfer 2008, pp. 198–199; also noted by Appell 1991)*

The harp portrayed at the beginning of this chapter reflects a common set of harp strings and relevant chords that create the personal songs plucked by Kenyah men. If we turn to the harp's frame, the elements of stability within the system, I see again the three sides: The first, shared with Bushler Bay, but manifested differently, is the ongoing oscillation between competition and cooperation. Whereas in the US, there's an underlying expectation of competition—that's what's emphasized, especially for men—among the Kenyah, cooperation is expected. Although cooperation is expected and common, the frequency with which men compete will also be clear in the pages that follow.

The second side of the Kenyah harp frame is the distinction between generosity and greed. Social pressure to share is phenomenally strong.[22] Yet there is recognition that people do behave selfishly. No one is surprised when a local leader cheats and takes more than his share, or when a family hides the meat they've brought home, keeping more for themselves—contrary to the norm of sharing all with their extended family and neighbours. But children are taught very consistently to share what they have, and all agree that's the right thing to do. I did not hear among the Kenyah the idea that there were ethical principles that one had to follow at great risk to one's self. They were eminently practical people.

The last side of the Kenyah harp pertains to the distinction between aristocrats (*Paren*) and commoners (*Panyen*) on the one hand, and a value on equality on the other. This distinction has come under attack, both from within and without. I describe later the headman, Pelibut's efforts to obliterate it; he brought both the Christian church and democratic ideals to bear in support of his view. People continued to consider the status of any man (or woman) according to this division, and it affected political, marital and economic decision-making. On the other hand, everything was somehow meant to be divided equally. Each person on a long expedition got an equal share, each person out hunting or fishing or collecting forest products got an equal share. Food in the household was divided into equal portions on all the plates. As with the other two sides of the harp, both concerns had to be taken into account.

Expedition-making as a masculine harp string

The activity that captured men's fancy and had traditionally taken a great deal of their time was expedition-making.[23] It was men's equivalent of swiddening for women. Kenyah men's 'harp strings' involved courage and making wondrous journeys; women sought to be hardworking and produce quantities of rice.

But some women went on journeys and most men participated in the agricultural cycle. Courage and hard work were admired in both men and women; the respective *emphasis* simply shifted somewhat.

In Long Ampung, sources of trade goods, salt and wage labour had been distant. So groups of men would routinely make long sojourns up over the central mountains of Kalimantan, fording rushing streams and making their way through dense forests to Sarawak, the closest access point for these desired goods. In 1980, men were still bringing back cloth, tools, kerosene, guns, cooking pots, tobacco, sugar, MSG and salt, among other things (Colfer 1985b).

Making such a journey was a dangerous undertaking, which required knowledge of paths through the forest, ways to garner foods from nature and care in protecting the group from dangerous animals or, in earlier times, other headhunters. Such trips were an informal rite of passage for boys, a study of needed knowledge for adult men and a test of leadership for organizers. They were also exciting adventures that many (though not all) men looked forward to and anticipated with both fear and delight. Bonds between men who went on expeditions together remained throughout life, as did memories of their adventures.

I was in Long Ampung in May 1980, when a group of 15 men and one woman[24] returned from one of these adventures.[25] They'd been gone a year. The community—excited both to see their menfolk and also the goods they brought back—immediately organized *Uman Usen* (to 'eat salt'), a ritual party to celebrate their safe return. Salt is essential to human life, and its unavailability locally meant that it was considered the gold of the Apo Kayan[26]—a key contribution men made to community health. By 1980, salt was more available; there was an airstrip a day's walk away at Long Sungai Barang, with a missionary plane occasionally stopping there. But salt was still in short supply. Bits of salt were given out to each family by the returning men, and salt was tasted as well. The men recounted their adventures, sharing new knowledge acquired in their travels. One time the men had brought back a new kind of pineapple plant that was not as spiny as that available locally. Another trip yielded the knowledge that an empty pressurized container could blow up. I was surprised at the variety of information, some quite esoteric (American kissing contests, trips to the moon), they'd gleaned. Kenyah curiosity and knowledge about the world beyond was impressive, particularly when compared with what I'd observed in rural Turkey in the 1950s, where villagers rarely left home and were more committed to obeying tradition.

BOX 4.1 EXPEDITIONS AMONG THE IBAN—THOSE LEFT BEHIND

In 1991–1992, I lived in remote Danau Sentarum (interior West Kalimantan) and conducted ethnographic research among the Iban and Melayu living there.[27] One day the Melayu headman of the village of Pulau Majang took me, reluctantly, by motorized canoe through the low brushy flooded forest that covered much of the region, into the beautiful, dense, high, flooded forest near the village of Empaik where I hoped to spend a few nights with the Iban. His anxiety at being in this forest and near these ex-headhunters was palpable—there was a long history of Iban taking Melayu heads more recently than Kenyah involvement in such pursuits. He tried to hide his fears and dropped me at the end of a trail with instructions about how to reach the village.

As I walked along the path, I came to a group of women sitting in a field hut, drinking *arak* (a strong alcoholic drink) as they took a break from their rice weeding. My appearance struck fear into their hearts—they had never seen a white woman before, and who knew what mischief I might be up to? The bravest of them came and asked me what I was doing, and as I explained that I'd just come to visit them, they gradually relaxed. My agreeing to sing some American songs helped to further dispel their anxiety. Eventually, after warning me that they would kill anyone who stole their husbands, they invited me to sit with them in the field hut and drink some of their *arak*.

As we chatted, it emerged that they were actually there consoling themselves. Their husbands had just left for an expedition to Malaysia. They expressed their fears that their husbands would find another woman and not return; though they also maintained that they were unconcerned about hidden affairs if they never learned about them. They expressed their sadness at not having their husbands' warm bodies on their sleeping mats (with extra giggles about implied sexual enjoyments) and that they would miss husbandly massages after a hard day's work in the fields. They knew they would be lonely (see Kedit 1991, on Iban circular migration).

Kenyah expedition-making continued, with the move to Long Segar, with important variations: Trips were shorter and now turned toward Indonesia instead of Malaysia. Rather than obtaining trade goods, men now worked for wages or under contract, mostly in forests, and brought back cash, chainsaws or outboard motors (Colfer and Dudley 1993). But the trips' functions as rites of passage, as adventures and as opportunities to develop and display leadership continued. Clever and ambitious Kenyah men became adept at speaking Indonesian and dealing with the many other ethnic groups intent on harvesting the province's wealth, outside investors with more money, power and education. The Kenyah had to learn to manoeuvre within the negative stereotypes (as 'primitive') that outsiders applied to them, but they were also able to make use of other groups' fear of them. The Kenyah headhunting history was known, and the idea that Dayaks had secret magic was also widespread.[28] Many of my colleagues from other parts of Indonesia openly expressed their fear of this group, whom some believed had tails. Once, in 1990, an Indonesian colleague, a forester, was so frightened of Dayaks that he feigned illness to avoid coming on a mandated field trip to Kalimantan (similar fears persisted in March 2019; See Chapter 7).

The leadership and politics harp strings

In Long Segar, Pelibut, the headman in whose home I lived, provides an interesting case of admired masculinity. The Kenyah change their names several times during their lives; the *Pe-* in Pelibut's name indicated that he had grandchildren. The *libut* was an adaptation of the Indonesian word for a thousand (*ribu*) and referred to his extensive expedition-making. He'd reputedly gone as far as Irian Jaya (now Papua) and had had marvelous exotic experiences throughout his life.

Among the Kenyah, traditionally there had been three social structural differentiations: aristocrats (*Paren*), commoners (*Panyen*) and in earlier days, slaves.[29] The *Paren* had had various rights vis-à-vis the *Panyen*. *Paren* had been able to use symbolic wooden carvings, certain named beads and artistic symbols on their elaborately beaded baby carriers (Whittier 1973) and graves. They had access to the community's labour, the right to lead and more.

Pelibut was born a commoner; he was the only commoner to become headman among all the Uma' Jalan Kenyah communities. He'd recounted to me his struggles to become leader, including being disdained and spat upon year after year in his earlier years. But he was a clever speaker, another admired masculine tone on the harp, and as his oratorical and intellectual abilities overcame community resistance to his leadership, he was able to lead the people first to accept Bungan Malan (a female deity who required adherence to fewer taboos than had the previous belief system), and then Christianity.[30] He had also fought for acceptance of democracy, which he'd argued required dropping the *Paren–Panyen* distinction. By 1979, it was locally 'illegal' to mention this distinction, subject to a fine (though its power still held sway).[31] He'd also led the group who initially settled Long Segar. Besides his leadership, expedition-making and oratorical skills, he was expert at the drawing of patterns for beadwork, carving in wood and bone, massage and blacksmithing; he readily admitted that he was no good at rice cultivation (the backbone of the economy and of women's role).

By the time I knew Pelibut, he was solidly ensconced as the leader of the community (*Ketua Adat*—customary leader), and no one overtly questioned his leadership. He called meetings and people came. When he negotiated contracts with outsiders involving community labour (e.g., to clear the landing strip across the river, to expand the church), people provided that labour. One time someone died the day I was to fly out from the Georgia Pacific Timber Concession to Samarinda. My colleague, Tamen Uyang, seriously disliked Pelibut's firm hold on power and obliquely questioned his right to lead. But in this instance, he told me that we could not go unless Pelibut agreed, as it was against their custom to leave the village when someone had just died. After some discussion, he agreed to ask for such permission. To my surprise, he did this in an overtly submissive way, clearly recognizing Pelibut's right to deny us permission.

This submission came about because of Pelibut's position in the community, but also because of his age. Unlike in the US, age among the Kenyah confers respect. Tamen Uyang was young, in his mid-30s; he was an aspiring leader and upwardly mobile. His mother's family was *Paren*, and his father's *Panyen*. As his mother had died at an early age, he'd been raised by her sisters, in a *Paren* household. However, his ambiguous status meant he had to prove himself worthy of a leadership position more strenuously than would have a full *Paren*.

In a discussion about the categorization of people, a group of men discussed their criteria. If a person always followed, and didn't give his opinion, then the elders didn't call him or ask him. If he agreed no matter what, they also didn't bother to ask his opinion; they were seeking people who could think. With *Paren pebeseq* (half aristocratic), if their children were not smart, their children reverted to *panyen*. To *alaq ngadan* (to 'get a name'; that is, to be publicly admired), a person must *njam pisiu* ('be able to speak'), and be *tigatawai* ('good hearted, friendly'). *Bek ia mpi tisen neng joo udip, mpi dulu mengin ia madung.* "If he doesn't know how to plan about life, people don't call him to sit [among the decision-makers]."

Tamen Uyang did not seek formal political position, claiming he did not wish to spend his time solving community disagreements or discussing community affairs at great length on his veranda. Instead he wanted to and did lead men on the Long Segar version of expeditions—contracting groups to log or clear[32] forests for money. He drew on his physical strength, linguistic ability and multicultural sophistication (another version of being 'clever at speaking') to negotiate contracts with outsiders. This sophistication was gained during his long and enforced stay in Malaysia during the 'Confrontation' between Indonesia and Malaysia (1963–1966).

Another man, Tamen Balan, had been appointed formal village leader. Tamen Balan was literate, making him an acceptable candidate to the government, and he was *Paren*, making him more acceptable to the community. However, he was young, mid-30s, which meant he could not comfortably challenge Pelibut, despite the latter's illiteracy. Nor did Tamen Balan exhibit any apparent wish to do so. When I asked why none of the other older *Paren* men had become leader, I was told that they were not clever at speaking, and thus could not lead effectively. *Paren* men readily acknowledged this shortcoming in themselves. All feared to debate Pelibut, whose leadership by then went unchallenged.

Relations among men were a bit reminiscent of the relations among loggers in the easy performance of a specified task. Men cooperated without much bossing behaviour;[33] each knew his (or a useful) part in performing the task. Attitudes toward time differed however. There was no concern about following any schedule; men didn't care how long the task took usually, and there was little interest in making its conduct efficient. 'Time' had not been 'money' traditionally. Nor had there been much interest in productivity. In Long Ampung, where most of the people I knew had grown up, their fields produced enough for the family with some to spare; a big surplus was of little advantage. In Long Segar, interest had grown in, and they had success at, producing surplus for sale, but attitudes about time were slow to change.

BOX 4.2 ON GOOD-NATURED KENYAH MEN

I was standing on the shore admiring the beauty of the Kayan River and the forests and fields beyond, in the comparative cool of late afternoon [in Long Ampung], just having finished my bath. A peaceful moment of pleasure. My attention was drawn to a raft coming around the bend in the river, with several men on it. It quickly became clear that this raft of logs was coming apart and that the men on board were trying to salvage the situation. They called out to those of us on shore, laughingly explaining their predicament. Several men near me jumped into the river, also laughing and working to retrieve the logs that had come free. With considerable effort, all good-humoured, they managed to reassemble the raft and continue down the river.

Images of American men working together kept coming to my mind. In the US, someone in the group would probably have been angry, blaming someone else for tying the logs incorrectly or for ineptitude in handling them or for slowness in retrieving them. The image of these men joyously solving life's problems remains with me, symbolizing the equanimity of the Uma' Jalan [Kenyah] in difficult situations and their ability to turn potential pain into pleasure (Colfer, Peluso, and Chin 1997, p. 7).

In one sense, the community was ruled by the old men. When a community decision had to be taken, the old men assembled on Pelibut's porch and discussed what ought to be done, usually far into the night. Eventually, a consensus (or nearly so) was reached and a decision was made. Anyone—man, woman or child—who wanted to listen and even participate in these discussions could do so, but the final decision-makers were these elderly men.

This same group was responsible for making legal decisions. When adultery occurred, for instance, the couple first tried to resolve it themselves. If that was ineffective, they involved their close family members. Men and women had equal voice in these small groups, influenced by the intelligence and wisdom of the individuals. If the problem was still not resolved, it was brought to these elderly men, who tried to understand what had happened, attribute blame and mete out appropriate punishment, normally a fine. Once the decision had been made and the fine paid, community members were prohibited from referring to it again. Anyone who was heard doing so was him/herself subject to a fine. Fines were typically shared between the victim and these elderly men. Being one of these elders involved more duties and hardships than raw power.

Courage as a harp string

Mashman (1993), drawing also on Freeman (1979), writes of the Iban:

> The underlying values of bravery (*berani*), boldness (*kempang*), and strength (*kering*) are central to any explanation of Iban aggression, warfare, and headhunting. A man of bravery and strength will have tattoos on his throat, a sign that he has endured pain, and a headhunter will have tattoos on his fingers . . . These values are reinforced through myth, ritual, and the daily division of labor.
>
> *(p. 236)*

The interest of Kenyah men in their own bravery has been briefly mentioned, but tattooing was not the avenue by which Kenyah men demonstrated it. It was Long Segar's women, rather than men, who sported tattoos; they were not symbolic of masculinity. Many elderly Kenyah women had a series of tattoos on their finger joints, lower arms and varying portions of their legs. Young women were unlikely to get tattoos even in the 1980s, but by the 1990s, some young men had begun to. At that time, the government was organizing the killing of tattooed men on Java, reportedly because of their assumed links to crime syndicates. This may have prompted a new (or renewed?) link with courage among Dayak men (though this is speculation).

The desire to prove one's courage takes on greater salience when we consider the practice of inserting pins into their penises. When I learned in February 1980—to my surprise—that many of the men of Long Segar indeed had their penises pierced, I asked why. I received the most common reasons given for their use in Borneo, also described by Brown (1991) and Zahorka (2017): the pleasure of a woman in intercourse and evidence of male courage.

A few prove their courage by climbing the glorious *tanyit* tree (*Koompassia excelsa*)—a tree reaching heights of 88 m (https://en.wikipedia.org/wiki/Koompassia_excelsa). The dramatic and beautiful *tanyit*, described earlier, was also often selected by bees for their honeycombs. I was told about groups of Kenyah gathering at the base of the tree, singing songs to encourage the bees to leave, while some brave man ventured up its trunk to the towering heights of its crown. Smoke was used to force the bees away, and the honey would be harvested and shared, to everyone's delight. Admiration was expressed for the courage, skill and danger such a harvest entailed.

The value on courage was not, however, absolute. A young man of our household turned over a tractor at work, injuring himself and others. When he came home, he was chastised by the older men for being overly *makang* (brave, fierce). Basically, he was warned against being foolhardy, told to admit when he didn't know how to do something.

Courage was also used in relation to speech. Pelibut referred once to a man sitting near us as not *berani* (brave) to take initiative. This man would only do what others told him to. The man himself agreed with this assessment. In another example, a man had talked of moving away from the Apo Kayan, a controversial topic, in front of a government official. His fellow villagers asked him why he had the courage to do that. Had the headman told him to? He responded, no, but the community had. I was told of young boys in the Apo Kayan who were *makang*, enduring the stings of pellets shot from bamboo 'guns' without flinching or running away, even when they drew blood. Tamen Uyang spoke of more physical competition among young men in the Apo Kayan in his youth, compared to the then-current young men. This could, however, have been the ubiquitous tendency to malign the youth of today, whatever the era. A man well known for his singing and composing ability was also admired for his courage: He recited one of his songs/poems (*dayong*) when the Indonesian soldiers were in Long Ampung during the Confrontation with Malaysia in the mid-1960s. It ended with the rhymed Kenyah phrase '*Jaat ale' se' alo' tantara*' ('Very bad looking, the foreign soldiers').

Women could also demonstrate courage however. The women told me that they did not cry out during childbirth. One told how she crossed her arms over her belly to give herself strength during contractions. In a 1980 survey of all the adult women in Long Segar, I asked whether the women spent the night alone at their field huts, typically far from the village—as a measure of marital trust/jealousy and autonomy. But responses also had a bearing on courage. Although some women expressed fear of killers and spirits as reasons for not spending the night there, 84% responded yes, when asked if they did so (Colfer 1985a, p. 202). The occasional involvement of women in expeditions was also recognized as courageous (as was my own travel from America to Indonesia). In Long Ampung, 12.5% of women had made such an expedition. In Long Segar, where 'civilization' is much closer, 50.2% had travelled (Colfer 1985b, p. 228).

Provisioning as a harp string

Not all expeditions were to distant places. People, especially men, went on shorter trips to hunt and fish to feed themselves, their families and neighbours. With some persistent wheedling, I persuaded Tamen Uyang to bring me along on a hunting and fishing trip with three other men (Tamen Long, his teenage son Madang and Tamen Uyang's uncle Tamen Kihin). We were heading up the Mela River, which split off from the Telen to the northwest, a few kilometres upriver from Long Segar, in two motorized canoes. Our destination was a densely forested and unpopulated area. The Mela, unlike the lazy, wide, brown Telen, was clear, fast-moving, and full of rapids. The trip upriver was exciting and beautiful, as we passed largely undisturbed areas of forest, interspersed with stretches of limestone cliffs. The further upriver we went, the narrower the river, the more rapids we had to traverse. Many times, we had to get out of the canoes and portage them and our belongings up over rapids to the next clear area.

The abundance of fish we were catching was amazing to me. But my very American excitement and glee were met with some discomfort. My fellow hunters told me that they'd believed in the past—and it seemed to me that in fact they still believed—that the fish were presenting themselves to us as a gift, that we should recognize the sacrifice the fish were making and accept it with dignity, gratitude and some solemnity. Their attitudes were protective and caring. I felt ashamed of my greed and thoughtlessness.

The men fished with cast nets—a sight I enjoyed, as these strong men draped the cast nets over their muscular arms and, with great skill and elan, tossed the net, creating a full circle of ripples on the water. Kenyah women did not find this sight of interest. For them, it was perhaps like an American woman seeing a man drive a car. To survive and thrive in the Long Segar environment, most young and middle-aged men were physically strong—or they had to come up with a niche role (discussed later in the chapter). Physical strength was appreciated in men for what they could thereby accomplish, and to that extent was associated with masculinity, but I never heard Kenyah men express any pride or admiration for their own or their friends' muscles per se (unlike the present-day Punan Murung of Central Kalimantan; Grossman 2017). Kenyah pride in physical strength had to do with its utility (including in competition), rather than masculinity per se. Strength and stamina in women were also admired.

Our fishers had accumulated many fish in the canoes by the time we reached a suitable camping area in late afternoon. There, two men began gathering poles to fashion a makeshift hut, complete with a pole floor for us to lie on, while the other two went off with their rifle to hunt. The fact that two men were good at hunting and two at fishing was accepted, with pleasure at the skills available to the group. Tamen Uyang, who had played an important role in fishing (but readily acknowledged his lack of skill at hunting), cooked the rice we had for dinner.

The hunters came back about dusk, and we began the arduous task of converting a deer into meat. Although I tried to be helpful, my lack of skills or knowledge was pitifully obvious. As night bore down on us, I was given the task of holding the flashlight while the men salted the catch. Utterly exhausted from a full day in the hot tropical sun and much physical exercise, I could hardly keep my eyes open. Yet no one suggested I take a rest, no special concern was expressed for me as a woman. I was expected to do my part, even if it was a very minimal, unskilled part.

When we finally were able to lie down, I expressed my fear about snakes possibly getting into my mosquito net. The men assured me they could (and would) take care of any snakes that came along. They expressed their own fear of *bali* (spirits). There was no shame in expressing such fears. It did not reflect badly on their courage; all Kenyah feared spirits. They considered fear to be logical and sensible in some situations. We made the practical arrangement that I would deal with the *bali* and they would deal with any snakes we encountered. All were happy with the trade.

The next day, returning to the village, we stopped a few kilometres short of Long Segar. The men took all the meat and fish out of the canoes and spread it on a blue plastic tarp on the river bank. They began dividing the harvest into five equal piles, one for each of us. I objected, saying that I had not contributed to the work, that I had been, if anything, a drain on their energies. But this reasoning was not accepted; indeed it prompted a rare expression of anger. The men insisted that I had come along, that even the smallest child who came along on such a trip would get an equal share. People were expected to do what they could (as had been shown the night before when, so exhausted, I was required to hold the flashlight); those who could contribute were expected to, those who couldn't were 'brought along' (*ilu ngkin ida*). Parents share with their children; children later share with their parents. The men all agreed about this, also making no gender differentiation. The Kenyah were proud that there were 'no Kenyah living under bridges', as some had seen in Java, and they felt that their strong ethic of sharing was an important reason.[34]

Readers may assume that men everywhere are providers. I do not believe that the Kenyah thought of men (any more than women) in that way. They spoke of men as providing money and goods; as contributing to rice cultivation, hunting, fishing and gathering forest products. But with women such a backbone of rice production,[35] and rice so strongly valued (as shown in the cognitive map in Figure 4.2, for instance), their image was of men and women contributing together to family prosperity and health.

Strength and protection as harp strings

In general, the denser the forest, the more relevant it was for men's subsistence activities and the more dangerous it was. Men tended to seek out new areas for rice fields each spring, in old growth (*mpa'*);[36] men hunted and gathered forest products that flourished in *mpa'*. The local rationale was that men were stronger, braver and more likely to be able to deal with dangerous animals (and in previous times, other headhunters). But this difference was one of degree. Women also collected forest products, fished, cared for gardens (*banit*), usually in secondary forest, and often made their way through dense forest to reach their rice fields; they, like the Javanese transmigrant women described by Elmhirst (2018), also valued their own strength.

The difficulties people encountered dealing with forest regrowth prompted me once to ask if people considered forests to be their enemies. Tamen Uyang first replied "yes", but as he considered it, he concluded instead that work was really their enemy, "because it kills us". I asked if people ever waged war (*pepatai*) against the forest itself. He replied, "Only if two people are fighting to see who's strongest, others might tell them to go fight with the forest and make a big rice field or carry a load of rice, if they want to show how strong they are".

One time, I went on an outing with several middle aged and older men, through old growth forest (*mpa'*) looking for a new area for the next year's rice field. My bush knife, which had been attached to my waist band, had fallen off and I hadn't noticed. Pebilung, a wizened man in his 60s who navigated the forest with amazing agility, told us to wait, that he would retrieve it. I doubted he could possibly find it, as we seemed to have been meandering through *mpa'* without any paths or guideposts. However, in about ten minutes, he returned, knife in hand. I was impressed by and praised Pebilung's agility and speed. All agreed, saying "Pebilung was very strong when he was young—indeed, they said if there were 1000 people, maybe there would be four like Pebilung when he was young . . . He could carry heavy loads and still go fast in the forest".

I made many trips with Tamen Uyang, both to other communities and to the many fields in the Long Segar vicinity. He did not overtly protect me. When we discovered that our canoe was directly under a poisonous snake relaxing on a branch once, he quickly moved the canoe—but that was protecting everyone in the canoe, including himself. He reminded me at night to take a flashlight when I wandered the village, as snakes also came out at night. But when my husband came and joined me there, Tamen Uyang reminded him as well. The women in my household also reminded me to bring my flashlight at night, not to go into the forest alone and to wear a protective sunhat when I left during the day. Men treated me as an equal, and that's also what I saw in most other male-female relations.[37]

There was one situation in which Tamen Uyang expressed dismay and fear relating to my safety. I don't know if this is evidence of male protectiveness of women, human protectiveness of humans or fear of adverse consequences. We were riding in a canoe that he was piloting, and he steered it a little close to the bank. A branch brushed our canoe and I avoided being hit only by brushing it away at the last minute. He was upset about his own handling of the canoe and my potential injury and asked me in a worried voice if my husband would not blame him if I'd been blinded by the twig. Those who harmed others, including unintentionally, tended to be fined. That applied to both men and women.

Headhunting and violence as harp strings

Headhunting, at least in the American imaginary, is linked to masculinity (Hoskins 1996). Kenyah men had once been headhunters—ending in the 1930s—and I imagined when I first arrived that I would encounter some related propensity for violence among them.[38] Pelibut was one of two people in the community who had any memory of actual headhunting, and that was when he was very young. In 1980, any propensity for violence seemed directed outward (like headhunting) toward other groups and internal relations were remarkably gentle and kind. I found no rite of passage ostensibly substituting for the role of headhunting in this process, as George (1996) found in Sulawesi.

Though the last head was discarded in 1963, people did talk about their headhunting past, sometimes with embarrassment, remembering wars with adjacent ethnic groups, particularly the Bakung. People reported hunting heads for various reasons: land disputes, fame, women and for religious purposes. Only the chief of the longhouse could hang heads in his rafters. The actual killer got a symbolic head to display. Tamen Uyang spoke of his childhood desire to headhunt—his grandfather had been a famous headhunter and community leader—and linked it with courage and strength. Pelibut remembered the excitement when men returned from a headhunting trip. He spoke of the community's earlier worries, when headhunting was forbidden and certain gender-differentiated taboos were abandoned, that men would become like women, a concern reiterated by Tamen Uyang.

Pelibut recounted earlier ideas about *sengka* (*Setaria Palmifolia*)—a long slender forest plant that became limp once cooked, which men avoided eating for fear of becoming 'woman-like' (*pekua' leto*). Some men still avoided *sengka*. Men also could not eat *payau* (sambar deer, *Cervus unicolor*), *pelanuk* (greater mouse deer, *Tragulus napu*) or *tela'o/uca'* (barking deer, *Muntiacus muntjac*) in times past. But such taboos were no longer observed.

Pelibut claimed to have been one of two men first to reject these ideas, when he accepted the deity, Bungan, a cult he maintained began in Long Ampung, in 1942 (Colfer and Dudley 1993; see also Urano 2010). By the mid-20th century, people still sacrificed pigs to her. They'd burn the hair on the back of the pig's body, the smell of which appealed to her, and they'd ask to be invincible in war and strong, rich and healthy in life. Then they'd kill and eat the pig. Their conversion to Christianity involved still more reduction in taboos.

There were a variety of links between headhunting and the supernatural. Supernatural help was regularly requested, and signs in nature were seen to grant protection to headhunters. One example involved waiting for a particular small bird to fly to the left of or through a circle hung between two trees to indicate that a headhunting outing would go well. Conversely, a hawk flying to the left was a bad sign, whereas flying to the right or in a circular pattern predicted a positive outcome. If the signs were right, people were said to have no fear.

I encountered very few violent incidents during my year-long residence: one, violence against malfeasant young teenagers; another, against a spouse; the third against a wife and child. In the first case, a big commotion out on Pelibut's veranda drew me out, early in my stay there. I saw Pelibut storming around the veranda, ranting and raving in a very threatening manner at two young teenage boys cowering in fear, tied to the support posts in the middle of the veranda. Although I did not as yet understand much Kenyah, others explained that Pelibut was threatening the boys with no food or water, with possible beatings and berating them for their actions. I was genuinely concerned about their welfare and struggled within myself with what I should do. The violence I'd anticipated seemed about to transpire.

It emerged that the boys had stolen a chicken from someone in the neighbouring Kutai community. Relations between the two communities, separated only by a small stream, were tense at the time. Both communities were part of a government resettlement programme, ostensibly sharing governmental inputs equally. But there were questions about the fairness of a recent division of plywood and roofing material. Additionally, the Kenyah were Christian, the Kutai Muslim; the two communities potentially competed for nearby lands for rice cultivation (a Kenyah specialty), for *beli'en* (*Eusideroxylon zwageri*) for making shingles (a Kutai specialty) and other forest products. Kenyah cows, which wandered free, had recently wrecked some Kutai gardens, raising the level of mutual antagonism (Colfer, Soedjito, and Azier 2008, orig. 1980).

I gradually learned that there was genuine concern among the men (and perhaps the women too) about the potential local political implications of this thievery and a feeling that the punishment had to be public and seen as serious by the Kutai. Little by little, as the day wore on, the boys' parents appeared and contributed additional lecturing and some dramatic posturing along with a few good whacks. Onlookers restrained parents who seemed in danger of injuring the young perpetrators. The ringleaders were singled out for more painful punishment, while the four younger boys, also tied to posts, looked on in fear. Adult men reminded them of their own culpability. Later that day, after additional pokes, prods, unattractive haircuts and further lectures by a policeman and a man from the nearby timber company, the boys were released, without much in the way of harm beyond the experience of fear, very unattractive haircuts and embarrassment. There was definitely more drama than abuse involved in this show, though it probably served as a potent lesson for the youths.

Beyond this, I never actually *saw* a Kenyah raise a hand to a child. Tsing (1996) concluded about the Meratus in Central Kalimantan that, although she heard a man threaten in public to beat his own child once (in proof of his courage), in fact she found "no evidence at all that child-beating ever occurred". The Kenyah are remarkably patient with their children, putting up with tantrums and other behaviour that few Americans would tolerate. However, the amount of social pressure that both parents and the community at large exert is considerable—on issues like generosity, cleanliness, respect for elders and fair contribution of labour. Expectations are high, but gentle means of enforcement are the norm for both men and women. Lying and thievery, though disapproved, are in fact more acceptable to Kenyah parents.

The second case also occurred early in my tenure there. Again, there was a commotion on the veranda. I found a group of people discussing a case of domestic abuse (Colfer, Peluso, and Chin 1997). One spouse had been beaten, was tired of being subjected to repeated attacks and had brought suit in the traditional way against the aggressor. The 'court' of elderly men had decided that the aggressor should be fined. The discussion at the time I arrived was about how much the fine should be. After some discussion, the men determined the amount of the fine and the division between the victim and the 'court' (as was usual). It was only some time later that I discovered, contrary to my initial assumption, that the victim was the *husband*, not the wife. Pronouns in neither Kenyah nor Indonesian specify gender. When asked why he hadn't fought back, my friend said, "He didn't like to fight". The woman was definitely seen in this case as the wrongdoer.

As time went on, I looked for further evidence of anything one might call child abuse; only one was uncovered. A man was reported to have attacked his wife and stepson with a beam. Over my years of involvement with this community, I heard of one more case of domestic abuse, by a woman. This woman, who attacked her husband with a wooden beam, was pushed to this extremity by his repeated infidelities and related public embarrassment. Violence within the community is not condoned for either sex.

I was told of one case of sexual violence that had occurred many years earlier. A group of young men were reported to have gang raped a girl who was mentally challenged in Long Ampung in the 1950s or 1960s. This was reported with shame and disapproval. And there were stories of inter-ethnic conflicts during the days of headhunting, when Kenyah women were reportedly raped by Bakung invaders. But accounts of this event vary; another version involves Kenyah women out hunting, being helped to carry home their pigs by Bakung men and making love with them of their own free will, thus angering their husbands who went to war with the Bakung. As in Indonesian, the term for force (*ase'*) is far more likely to refer to social pressure or unwanted verbal advances than to physical violence.

Sexuality as harp string

> Everything in this world has male and female.
> There can't be man without woman or woman without man.
>
> (Pejalong, a man, an elder, 1980)

> First, we want to talk with each other (across the sexes), then we want to touch, then we want to sleep together, then we want to make love. That's true of all human beings, men and women. And if we begin to feel that things are going in that direction, the proper thing to do is to tell your parents or your relatives, or just the people of the community so they can arrange a marriage. Don't just go sleep together.
>
> (Tamen Uyang, a middle-aged man, 1980)

> Young boys have a thing [penis] that gets angry [*a'un ca inu neng aang ilu ya' keto'*] and wants to live [*mudip*] if it is near girls.
>
> (Pelibut, a man, an elder, 1980)

Attitudes toward heterosexuality initially reminded me of the situation in US high schools in the 1950s. There was mild disapproval of premarital sexuality, greater disapproval of extramarital sexuality, and in both cases, such disapproval was greater for women than for men, though both clearly engaged regularly. There was also an element of secrecy about it. But any such tendencies were strongly reinforced by Christian missionaries, who had a significant impact on local beliefs.

Pelibut, for instance, had married a number of times in Long Ampung before settling down with his final wife. Earlier in his life, marriages had been easily arranged and dissolved without fanfare or disapproval. But the Protestant KINGMI church (a Lutheran offshoot) forbade divorce, whereas interestingly, the local Catholic church did not.

When I first arrived in Long Segar, I had to report to the local *Camat* (county head), a Javanese man, appointed by the Indonesian government. Discussions with him revealed that he had sexual access to some young Dayak women within his *kecamatan* (county)—access that was accepted, somewhat reluctantly, by Long

Segar community members. My sense was that outsiders who engaged sexually with Kenyah women tended to feel that such behaviour was an affront to Kenyah men, a sort of one-ups-manship, but I saw no evidence that Kenyah men saw it that way. Kenyah men did not control women's sexuality, nor were women considered to *belong to* their husbands or fathers. Women were locally expected to make their own decisions about their sexuality (also noted by Tsing 1993, among the Meratus)—though the community might not be particularly happy about those decisions, some of which resulted in unplanned pregnancy.

Tamen Uyang advised one woman whose husband was involved in an extramarital affair not to leave as the husband would be happy, having a 'standby booking' (his words!) and she'd have a hard time finding another husband, with child in hand. On another occasion he said, "men can close their eyes, and there are lots of women to choose from, but women can't find another man even if they open their eyes wide searching [with appropriate kinesics]". There are also proverbs about this inequity:

1. *Ia tusa pelenca urong*—lit. it's hard to make a gourd sink. Fig: It's hard for a woman to find a man.
2. *Ia kimet ulu re, yare' uben iko' oo'*—lit. She thought it was the head but really it was the tail (of the snake). Fig: She thought he was the first in a long string, but really he was the last to ask for her hand.

On the other hand, a joke told about both men and women seeking divorce suggests equality:

> While discussing divorce, a couple went to their rice storage hut. The husband went up to get some rice and the wife stood below. She saw her husband's penis and decided against the divorce. They went home and made up.

The same story is told with a man standing below looking up at his wife's vagina, with the same outcome.

Men are usually expected to be the initiators of courtship. As with the Iban, young men would ask (*menyat jaboq*) and be given permission to sneak into a girl's sleeping mat and lie quietly, snuggling with her. Such affairs were conducted secretly, and were accepted, even expected, with the idea that intercourse was not to take place (though it sometimes did). When my late-adolescent son came for a brief visit, local girls attempted to seduce him, after which local men mildly reprimanded him for acquiescing. Such sleeping together suggests an ongoing relationship.

More powerful Kenyah men locally had easier sexual access to women than did others. Two middle-aged leaders I knew well were involved in multiple extramarital affairs.[39] When/if they were caught, they were required to pay a fine; if they persisted, they were threatened with expulsion from the community, though I did not know of any cases where this actually happened. The fact that husbands and wives shared their incomes meant that in effect the spousal victim was doubly impacted when a fine was levied: by the initial adultery and by losing some of the household's income/assets. Women could also be fined for adultery—most commonly considered among widows or divorcées.

There was no American-style 'war of the sexes'.[40] Where Americans tell tales of the Amazons fighting against men, the Kenyah tell myths about magical women coming to the rescue of Kenyah heroes (e.g., Awing in the tale of Balan Tempau).[41] The situation coming closest to such 'warfare' occurred when groups of men and women worked together in rice fields. Young men and women took turns singing; the gist of the singing was cross-sex teasing about who initiated sexual encounters.

Another idea, which could mildly pit men against women relates to ideas about semen. Some men expressed the idea that making love cumulatively reduced men's lifelong supply of semen and ultimately drained their strength.[42] This served to discourage sexual excesses for some men.

Whereas many American men expressed admiration for a man who makes love with many women, the Kenyah expressed ambivalence. One man was described, with humor and a touch of disapproval, as being *salet sait* ["strong at fucking"], meaning that he and his wife made frequent, energetic and prolonged love. Indeed, this was mentioned in connection with purported fear that their lovemaking would bring down the field hut (on stilts) where two couples were staying. Another man got angry when his men friends called out publicly that he *sait* [was fucking] a woman he was not married to, even though he was in fact doing so. He wasn't worried about being fined, but rather the disrespect. The dual purpose of penis piercing has already been mentioned: demonstration of courage and enhancing a woman's sexual pleasure.[43]

Although I did not know of any individuals whose sexual orientation was different enough to remark on, I did have a couple of conversations suggesting that at least the older generation was aware of possibly non-heterosexual alternatives. Pelibut described two men in the Apo Kayan who had done everything like women. They dressed as women, did the work of women, and in get-togethers joined the women. They never married and lived alone. He expressed no rancor or sanction, only amused acceptance. This came up because a visitor's little boy looked like a girl. He had worn skirts for a long time and still (age six) refused to have his hair cut. The boy listened throughout this conversation, in which no disapproval was shown—again, merely amused acceptance. He hid his face in his mother's lap but seemed shy rather than ashamed. I recorded another similar instance, of a five-year-old boy wearing a dress, whose behaviour was commented on, a simple observation without apparent disapproval.[44]

I had been told that whistling was inappropriate for women, but forgot one day:

> I apologized for forgetting and whistling and Pelibut, who served *in loco parentis*, reiterated, "Don't whistle. A woman [who does this] imitates/acts

> like a man" [*Ayen nyepeleu, leto pekua' laki*]. That's what they tell children. The word *kelake'* means a woman who does men's work. She cuts wood, makes *sirap* [shingles] and *prahus* [canoes]. *Kelake'* marry and otherwise lead normal lives. He said I was one. But for men, *keledo'*—men who do women's work—they don't marry or have kids, are afraid of the dark, cook well, like to play with[45] the youth [*pemuda*], take care of children and pigs. His adult daughter and the *pemuda pemudi* [youth] were sitting listening amusedly. They didn't know about this either but thought it was funny. There were two [like this] when they were in the Apo Kayan [Long Ampung].
>
> *(notes, Long Segar, 1980)*

As with many things, harmless difference is noted. I did not see it as real disapproval at the time. But insofar as it may have been, it is far less extreme than the aversion one saw in those days, and in some places still, in the US or now in parts of Africa.

The arts as harp strings

I mentioned earlier the prestige options for men who did not excel in physical strength, clever speaking, expedition-making or even bravery. There was an appreciation for the utility of various talents and a willingness to seek out the appropriate niche for any non-conforming man. I'm reminded of one young man who was obviously neither strong nor particularly coordinated. He was not good at sports—soccer being played every Sunday afternoon.[46] His family encouraged him to excel at school, to aspire to teach school (as he in fact did), since it was clear that he would not excel at the traditional masculine roles/skills.

Other men, not as obviously deficient in these realms, were also able to excel in different arenas. One man was skilled at playing the *sampe'* (a guitar-like instrument). He also had an excellent memory for Kenyah traditional myths and stories, able to delight his audiences. One time, at a good-bye party for me, he entertained us all by creating a series of refrains to a song he'd produced that poked friendly fun at my interview questions, which asked over and over about numbers of trees, plants and fields: *Koda' kadu'? Koda kadu'?* [how many? How many?].

Other men were excellent dancers. They danced, holding a shield, with an elaborate headdress of tall hornbill feathers. They could swirl around, bent low, looking up into or hiding behind imaginary trees, imitating or hunting various animals, seeking out headhunters (*ayau*) to engage in battle using elaborately carved swords. When dance parties were underway, young girls would typically go and pull men of all ages to their feet, one by one, insisting they dance for the crowd. Although women's dances were different, they too were pulled onto 'centre stage' as often as men. Most people complied, though with varying degrees of reluctance, enthusiasm and skill.

Older men, beyond expedition-making or carrying heavy weights, became skilled at weaving fish traps or splitting rattan or bamboo for use in baskets.

These skills were appreciated, but old women also spent time making valued handicrafts (weaving baskets, fishnets, mats). Any masculine weighting was muted, if it existed at all.

Another case of an unusual but admired man related to the healing arts. Women had traditionally been the midwives, but this man—who had lost half of one leg in a forest encounter with a wild pig—had somehow demonstrated his ability in this realm, and was unanimously accepted as a skilled birth attendant, helping many women through their births. Some women were helped by their husbands as well.

Dominant masculinities summarized

As with the men in my own personal life and in Bushler Bay, masculinities in both Denpasar and Long Segar included valued masculine characteristics, interests and norms/roles. But both Indonesian cases stand out in the muted nature of gender differentiation. This muting was much more extreme in Long Segar, summarized here.

Ideology

The Kenyah (unlike the Balinese) exhibited no gendered ideology, despite being bombarded with that of the Christian church, to some extent the Indonesian government and the negative stereotypes of other outsiders (see e.g., Dove 1999, on planters in Indonesia). The idea that women were inferior or *inherently different* was absent.

Characteristics

Men were considered likely to be stronger, more courageous and better at speaking publicly. But many women also exhibited and admired these qualities in themselves.

Interests

Men were most likely to express interest in travel, adventure, leadership and hunting. Although men were the operators of heavy equipment (for timber companies), they did not express the fascination I saw among American men with such vehicles. Kenyah men's interest (like their interest in the utility of their own physical strength) was linked to the capability of such equipment to reduce community labour rather than to inherently masculine qualities.

Gender roles

Kenyah gender roles were unusually flexible, with the exception of politics. Though women and youth were free to participate in political discussions if they chose, they rarely did so once an issue reached the community level. There, old men took charge. Men of middle and old age dominated in interactions with outsiders as well.

Men tended to make expeditions in search of adventure, money and goods. They tended to pilot canoes and handle chainsaws (due to their greater physical strength). They cut the trees to clear swiddens and they carried extra heavy loads. But women also often piloted canoes, traversed forests, gathered forest products and carried home backpacks full of harvests; and men often washed clothes, took care of children and cooked—some of women's more usual tasks—as needed.

Kenyah men in forests: conclusions

The depth of the links between Kenyah men and the forest at that time are difficult to convey. The Kenyah lifeway was dependent on the forest in almost every way. Their rice came from swiddens cut from the forest by Kenyah men; many of their dietary complements (fruits, vegetables, leaves) came from the varying stages of forest regrowth (old swiddens and the primary forest where men felt at ease). Their ways to demonstrate skill and acumen were via hunting and fishing in the rivers, many of which were kept clear by the existence of the surrounding forests. The beautiful handicrafts men made from rattan or bamboo or wood were dependent on the forests which provided the raw materials. Kenyah men's dancing mimicked the animals and activities that characterized their forest-based life. A crucial traditional way to achieve distinction derived from making expeditions into and through the forests. And their ethnic identity was forest-based (see Tsing 1999, on Central Kalimantan, or Elmhirst and Darmastuti 2015, on Lampung, Sumatra, for similar evidence of forest-based ways of life).

Of course, these men also played a crucial role in obliterating the forest, as part of their breadwinning norms. Their excellence at speech granted them access to the wealth and power of outsiders, whether from government or industry. Their physical strength and courage, along with their desire to demonstrate both, contributed to their willingness and capability to clear the land for transmigration projects, or industrial timber and oil palm plantations. Although less adventurous than a journey to Malaysia, travel within Kalimantan on such contract bases made satisfactory substitutes for their traditional expedition-making. It also allowed them to visit their families more often, contribute labour at home when needed, and still 'bring home the bacon'.

To write these words at this time, remembering so well and knowing that this habitat and aspects of this culture are truly gone, pains me. But as I discuss in Chapter 7, the Kenyah are a remarkably adaptable and resilient people—including their gender system. These men, who are able to cry in public without embarrassment (Chapter 6), understand and cry with me; but they are carrying on with the courage, hard work and leadership they value in themselves.

On this journey, we took a side trip here into urban and elite Balinese masculinities, and a fuller stop with a group that anthropologists used to call 'primitive', the Kenyah. In the next chapter, we move to two cultures once called 'peasant', the Javanese and the Minangkabau; from a forest-based culture to agriculturally-based systems.

Notes

1 Gender differentiation has increased in urban Indonesia, along with the strength of Islamic fundamentalism (supported internationally) and the spread of the gender policies and stereotypes that characterized Soeharto's New Order.
2 The material on Bali is drawn from several hundred pages of fieldnotes/journal entries (20 June to 21 August 1979) and from memories, accrued during a study on family planning for my master's in public health from the University of Hawaii.
3 The information on forests is drawn from Whitten, Soeriaatmadja, and Afiff (1996).
4 This gentleness in daily life goes hand in hand with extraordinary violence that occurred there in 1965, violence that was not discussed then in Bali and about which I only slowly learned over the years. A similar conflict struck me between the smooth and gentle social relations among the Kenyah, who were reputed to have been fierce headhunters.
5 Besides his wife and daughter, a varying number of servants, mostly relatives from his village, lived there.
6 Thanks to Rebecca Elmhirst for reminding me of these associations.
7 I subsequently examined this claim in Sitiung, West Sumatra (Chapter 5). The word, *paksa*, usually translated as 'force', very often also means serious social pressure. Social pressure from the village leader and other officials was very effective among the Javanese who are very comfortable with their hierarchical social system. This was the kind of 'force' applied there.
8 Much as I decry this attitude and welcome its near-disappearance in Indonesia now, I personally benefitted greatly from it. It was instrumental in overcoming my fear of public speaking, such that I learned to share my own points of view much more freely.
9 Though certainly men have serious advantages in terms of voice/control within the household, acceptability of polygyny, rights to children in case of divorce and comparative freedom to engage in extramarital sex.
10 He may even have been perceived as *more* qualified, if one recognizes the advantages of having a quantitatively oriented profession (demography). On the other hand, being Australian, he may have suffered from association with the general, local distaste for the Australian surfer world in Bali.
11 Though from an American perspective I was pitifully poor at the time.
12 The material for this section comes from the following: published material (Colfer and Dudley 1993; Colfer, Peluso, and Chin 1997 and Colfer 2008); one year of co-residence with Tamen Uyang, field assistant and ultimately one-time lover and father of my son; six large plastic boxes of fieldnotes from many repeat visits and survey results; my own memory and interpretations; personal journals.
13 This description draws on material in MacKinnon et al. (1996).
14 Korean and Malaysian companies operated adjacent to the Georgia Pacific concession.
15 The fear that people from Java tended to have of Dayaks also contributed to their reluctance to stray far from base camps (see Chapter 7).
16 Ardener (1975) provides numerous examples of the muting of women's voices, but here I refer instead to a more general muting of a whole axis of differentiation, glimpsed initially and less completely in Bali, one that has great salience in other contexts. Interesting in light of Connell's (1995) view: "A culture which does not treat women and men as bearers of polarized character types, at least in principle, does not have a concept of masculinity in the sense of modern European/American culture" (p. 68). Connell also notes the muting of gender differences for one of the Australian environmentalists whose life histories she took (p. 138). In later work, Connell (2005) also argues that "cultural consent, discursive centrality, institutionalization,

and the marginalization or delegitimation of alternatives are widely documented features of socially dominant masculinities" (p. 846), none of which were evident among the Kenyah.

17 Other relevant analyses of Indonesian groups around that time include, e.g., Dove (1981), Drake (1991), Mashman (1993), Tsing (1990), Colfer et al. (2001); the collection by Ong and Peletz (1995) and others. More recently, similar ideas seem to hold: e.g., Elmhirst, Siscawati, and Colfer (2016), Elmhirst (2011), Haug (2017) and Li (2015).

18 The concepts came from local conversation, in the Kenyah language. From Long Segar's population of about 1000, 49 adults responded to the instrument (see Colfer 1982). Sampling took advantage of naturally occurring groups: 30 men working on a village project; and 19 women, on a church project. A similar study is discussed in Chapter 5.

19 The Kenyah word for this was *man inu*, which means 'making things'. It referred to making traps, mats, baskets, beaded panels and jewelry, fishing nets and other household items—all of which were typically made for subsistence or local use rather than sale. I only saw men making traps and women making beaded panels and jewelry, but I also never heard the idea that only one gender or the other could make any of these items.

20 What I describe here reflects the situation in the 1970s–1980s. After that time, there was increasing land pressure on Long Segar, due to the arrival of timber plantations, transmigration and now oil palm plantations. These people were by 2016 intimately involved also with oil palm as a commodity (Elmhirst, Siscawati, and Colfer 2016). Chapter 7 provides a 2019 update.

21 In an observational time allocation study conducted in 1979–1980, 30% of women's time was spent in agriculture, vis-à-vis 26% of men's (Colfer 1981, p. 81).

22 I remember my attempts to keep a few chocolate bars I'd brought from the city for my own selfish enjoyment, sneaking them under cover of darkness, knowing if found, I'd have to share them all.

23 We find some similar interests from Britain, which

> bequeathed its English-speaking colonies narratives in literature that merge conceptions of idealized masculinity with leadership and heroism. Many of these narratives revolved around the concept of the journey and the quest.
>
> *(Hall 2011, p. 33)*

Whereas leadership and journeys were key, western-style heroism and quests were not emphasized among the Kenyah.

24 Tsing (1993) also talks about some Meratus Dayak women's adventurous spirit.

25 In April 1981, my assistant, Tamen Uyang, returned to Long Ampung for further data collection and found 50 men away on expedition.

26 The Apo Kayan [Kayan Highlands] is the area near the Malaysian border in which Long Ampung is located. The Kayan is a river.

27 As noted by Giesen and Aglionby (2000), "DSNP [now Danau Sentarum National Park] consists of a series of interconnected seasonal lakes (*danau*), interspersed with swamp forest, peat swamp forest, and dry lowland forest on isolated hills" (p. 5). In 1991, it was 80,000 ha and considered a government *Suaka Margasatwa* (wildlife reserve). The population of 6500 fisher folk lived in 39 villages in or near the park. Forests were well used for timber and non-timber forest products. See e.g., Colfer et al. (1997, 1999, 2000) or Colfer (2006).

28 Indeed, before I went to the field, I was told by another anthropologist that Margaret Mead had told him that Dayaks had potent magic, that he should be careful of what he drank from their hands. An urban legend?

29 Another differentiation that was important in Long Segar was between Protestants and Catholics. Each group had its own neighbourhood and church, with most *Paren* associated with the Catholics (who'd come later to Long Segar, beginning in 1972).

30 McKay and Lucero-Prisno (2012) summarize masculine concerns across Southeast Asia, including "ideal male qualities of *malakas* or strength, embodied in pre-colonial *datu* chiefs or 'Big Men' who combined bravery, physical strength, intelligence, eloquence and rapport with the spirit world to gain followers" (p. 23, italics in original)

31 One *Paren* woman, unhappy with Pelibut's leadership, likened the *Paren* to the valued ironwood (*beli'en, Eusideroxylon zwageri*), which was being strangled by the [*Panyen*] strangler fig (*lunok, Ficus* spp.).
32 E.g., his subsequent "large contract to clear forest for *kelapa sawit* [oil palm] with [a British plantation manager], from TAD's plantation at Muara Wahau" (journal, 28 March 1996).
33 Cf. the hierarchical organization displayed in the Bushler Bay basketball games, or the obedience required in many American bureaucracies.
34 The strength of this ethic for both men and women is hard to convey. Any excess was expected to be shared. My ten-year-old daughter brought one Barbie doll with its multitudinous accoutrements with her in the one suitcase she was able to bring, her only material links with home. When I did not insist she share out these items, the disapproval was tangible. Many conflicts and antagonisms within the community derived from this ideal of sharing everything equally.
35 In Mozambique, men's breadwinning also loses significance, though the emphasis on men's sexuality is more extreme than among the Kenyah:

> In a country where women represent a major part of the labor force, namely in agriculture, breadwinning responsibilities are almost unimportant for an understanding of symbolic gender differentiation when compared with the key role played by sexuality in reproducing the standards of male dominance.
>
> *(Aboim 2009, p. 216)*

36 Although much of what I write remains today, there is no old growth left in accessible areas (see Chapter 7).
37 There are two exceptions to this pattern: (1) an individual lacking in intelligence: Such people were treated in a less egalitarian way; and (2) in the public, political world: Although women spoke routinely in small groups and within families, and they could participate in public discussions if they chose, they rarely did so. They believed themselves and were seen by others as not typically 'clever at speaking'.
38 Indeed, my parents expressed fears that my head would wind up on a Kenyah veranda.
39 There were other leaders who appeared to remain faithful to their wives.
40 Appell's (1991) description of the Rungus seems applicable to the Kenyah as well:

> Among the Rungus [another Bornean group] there is no evidence that females envy males, or vice versa, and there is no evidence of an underlying layer of aggression or antagonism with respect to the opposite sex as represented in mockery, jokes, overt statements, or the play of children. Boys do not tease girls or belittle female roles, and girls in their play do not tease boys or ridicule any of the male roles. Nor is there any association of aggression in coitus in terms of bodily injury either in cases of fornication, adultery, or marital intercourse.
>
> *(p. 86)*

41 Cf. Nyawalo's (2011) Kenyan [African] tales of womanly betrayal:

> The vulnerability of a man's life in the hands of a woman is portrayed . . . In the story (of Luanda Magere, 'a fierce and brave warrior' of the past] . . . it is the wife who manages to destroy a man whose strength was legendary among his fellow men.
>
> *(p. 128)*

42 Keeler (2017) refers to such beliefs in South Asia and their absence in Burma (p. 224).
43 See also McKay and Lucero-Prisno III (2012) for a brief discussion of the practice among Filipino seamen of inserting small balls along the penis shaft, reported to be 'something extra' for partners (pp. 28–29).
44 I saw the same acceptance of boys behaving more like girls in 1991 in West Kalimantan, where one teenage Melayu boy regularly wore lipstick for a time. It wasn't even remarked upon.

45 In Kenyah, 'to play with' (*main*) can be sexual or not, as in English. I did not think he was implying sexuality here, but he could have been.
46 Very different from the basketball games in Bushler Bay. Whoever wanted to play joined either the youth (*demanai*) or the fathers (*tamen*) team; the audience came and went as the spirit moved them. No one seemed to care much who won, nor was adherence to rules emphasized, though expertise in playing was admired. Sports competitions between communities were, however, taken more seriously.

References

Aboim, Sofia. 2009. "Men between Worlds: Changing Masculinities in Urban Maputo." *Men and Masculinities* 12 (2):201–224.

Appell, George N. 1991. "Individuation of the Drives of Sex and Aggression in the Linguistic and Behavioral Repertoire of the Rungus." In *Female and Male in Borneo: Contributions and Challenges to Gender Studies*, edited by Vinson H. Sutlive and G. N. Appell, 57–120. Williamsburg, VA: Borneo Research Council.

Ardener, Shirley, ed. 1975. *Perceiving Women*. London: Malaby Press.

Brown, Donald E. 1991. "The Penis Pin: An Unsolved Problem in the Relations Between the Sexes in Borneo." In *Female and Male in Borneo: Contributions and Challenges to Gender Studies*, edited by Vinson H. Sutlive and G. N. Appell, 435–454. Williamsburg, VA: Borneo Research Council.

Colfer, Carol J. Pierce. 1981. "Women, Men and Time in the Forests of East Kalimantan." *Borneo Research Bulletin* 13 (2):75–84.

Colfer, Carol J. Pierce. 1982. "Women of the Forest: An Indonesia Example (republished as 'Uma' Jalan Perceptions of People and Forest' in Colfer 2008. Longhouse of the Tarsier. pp. 243–263)." In *Women in Natural Resources: An International Perspective*, edited by M. Stock, J. Force, and D. Ehrenreich, 153–182. Moscow, ID: University of Idaho Press.

Colfer, Carol J. Pierce. 1985a. "Female Status and Action in Two Dayak Communities." In *Women in Asia and the Pacific: Toward an East-West Dialogue*, edited by M. J. Goodman, 183–214. Honolulu: University of Hawaii Press.

Colfer, Carol J. Pierce. 1985b. "On Circular Migration: From the Distaff Side." In *Labour Circulation and the Labour Process*, edited by G. Standing, 219–251. London: Croom Helm.

Colfer, Carol J. Pierce. 2006. *Vignettes from a Year in Borneo: Local People and Conservation*. Lulu.com.

Colfer, Carol J. Pierce. 2008. *The Longhouse of the Tarsier: Changing Landscapes, Gender and Well Being in Borneo*. Edited by G. N. Appell and Vinson H. Sutlive. Vol. 10, *Borneo Research Council Monograph Series*. Phillips, ME: Borneo Research Council, in cooperation with CIFOR and UNESCO.

Colfer, Carol J. Pierce, and Richard G. Dudley. 1993. *Shifting Cultivators of Indonesia: Managers or Marauders of the Forest? Rice Production and Forest Use Among the Uma' Jalan of East Kalimantan, Community Forestry Case Study Series 6*. Rome: Food and Agriculture Organization of the United Nations.

Colfer, Carol J. Pierce, Nancy Lee Peluso, and See Chung Chin. 1997. *Beyond Slash and Burn: Building on Indigenous Management of Borneo's Tropical Rain Forests, Advances in Economic Botany; v. 11*. Bronx, NY: New York Botanical Garden.

Colfer, Carol J. Pierce, Herwasono Soedjito, and Albar Azier. 2008 (orig. 1980). "On Resettlement: From the Bottom Up." In *The Longhouse of the Tarsier: Changing Landscapes, Gender and Well Being in Borneo*, edited by Carol J. Pierce Colfer, 45–62. Phillips, ME: Borneo Research Council, UNESCO, and CIFOR.

Colfer, Carol J. Pierce, R. L. Wadley, A. Salim, and R. G. Dudley. 2000. "Understanding Patterns of Resource Use and Consumption: A Prelude to Co-management." *Borneo Research Bulletin* 31:29–88.

Colfer, Carol J. Pierce, R. L. Wadley, and P. Venkateswarlu. 1999. "Understanding Local People's Use of Time: A Precondition for Good Co-management." *Environmental Conservation* 26:41–52.

Colfer, Carol J. Pierce, R. L. Wadley, J. Woelfel, and E. Harwell. 1997. "From Heartwood to Bark in Indonesia: Gender and Sustainable Forest Management." *Women in Natural Resources* 18 (4):7–14.

Colfer, Carol J. Pierce, Joe Woelfel, Reed L. Wadley, and Emily Harwell. 2001. "Assessing People's Perceptions of Forests: Research in West Kalimantan, Indonesia." In *People Managing Forests: The Links Between Human Well-Being and Forests*, edited by Carol J. Pierce Colfer and Yvonne Byron, 135–154. Washington, DC: Resources for the Future.

Connell, Raewyn W. 1995. *Masculinity*. Berkeley, CA: University of California Press.

Connell, Raewyn W. 2005. "Hegemonic Masculinity: Rethinking the Concept." *Gender and Society* 19 (6):829–859. https://doi.org/10.1177/0891243205278639.

Directorate General of Forest Utilization. 1989. *Report on Field Case Studies of Forest Concessions: Appendix 5 on PT Kiani Lestari, East Kalimantan*. Jakarta, Indonesia: Ministry of Forestry, Government of Indonesia and FAO.

Dove, Michael R. 1981. "Subsistence Strategies in Rain Forest Swidden Agriculture: The Kantu' at Tikul Batu." PhD, Anthropology, Stanford University.

Dove, Michael R. 1999. "Representations of the 'Other' by Others: The Ethnographic Challenge Posed by Planters' Views of Peasants in Indonesia." In *Transforming the Indonesian Uplands: Marginality, Power and Production*, edited by Tania Murray Li, 203–229. Amsterdam, Netherlands: Harwood Academic Publishers.

Drake, R. A. 1991. "The Cultural Logic of Textile Weaving Practices among the Ibanic People." In *Female and Male in Borneo: Contributions and Challenges to Gender Studies*, edited by Vinson H. Sutlive, 271–294. Williamsburg, VA: Borneo Research Council.

Elmhirst, Rebecca. 2011. "Migrant Pathways to Resource Access in Lampung's Political Forest: Gender, Citizenship and Creative Conjugality." *Geoforum* 42:173–183.

Elmhirst, Rebecca. 2018. "Departing from Java to Lampung: Locating Javanese Diasporic Practices in Indonesia's Transmigration Program." In *Departing From Java: Javanese Labour, Migration and Diaspora*, edited by Rosemarijn Hoefte and Peter Meel. Copenhagen: NIAS Press.

Elmhirst, Rebecca, and Ari Darmastuti. 2015. "Material Feminism and Multi-local Political Ecologies: Rethinking Gender and Nature in Lampung, Indonesia." In *Gendered Entanglements. Revisiting Gender in Rapidly Changing Asia*, edited by Ragnhild Lund, Philippe Doneys, and Bernadette Resurreccion, 177–206. Copenhagen, Denmark: NIAS Press.

Elmhirst, Rebecca, Mia Siscawati, and Carol J. Pierce Colfer. 2016. "Revisiting Gender and Forestry in Long Segar, East Kalimantan: Oil Palm and Divided Aspirations." In *Gender and Forestry: Climate Change, Tenure, Value Chains, and Emerging Issues*, edited by Carol J. Pierce Colfer, Bimbika Sijapati Basnett, and Marléne Elias, 300–317. London: Earthscan/CIFOR.

Geertz, Clifford. 1963. *Agricultural Involution: The Processes of Agricultural Change in Indonesia*. Berkeley, CA: University of California Press.

George, Kenneth M. 1996. "Lyric, History, and Allegory, or the End of Headhunting Ritual in Upland Sulawesi." In *Headhunting and the Social Imagination in Southeast Asia*, edited by Janet Hoskins, 50–89. Stanford, CA: Stanford University Press.

Giesen, Wim, and Julia Aglionby. 2000. "Introduction to Danau Sentarum National Park, West Kalimantan, Indonesia." *Borneo Research Bulletin* 31:5–28.

Grossman, Kristina. 2017. "Gaharu King—Family Queen: Material Gendered Political Ecology of the Eaglewood Boom in Kalimantan, Indonesia." *The Journal of Peasant Studies* 44 (6):1275–1292. https://doi.org/10.1080/03066150.2017.1341408.

Hall, Maurice. 2011. "Negotiating Jamaican Masculinities." In *Global Masculinities and Manhood*, edited by Ronald L. Jackson II and Murali Balaji, 31–51. Urbana, IL: University of Illinois Press.

Haug, Michaela. 2017. "Men, Women, and Environmental Change in Indonesia: The Gendered Face of Development among the Dayak Benuaq." *Austrian Journal of South-East Asian Studies* 10 (1):29–46.

Hoskins, Janet. 1996. *Headhunting and the Social Imagination in Southeast Asia*. Stanford, CA: Stanford University Press.

Kedit, Peter M. 1991. "'Meanwhile, Back Home . . .': Bejalai and Their Effects on Iban Men and Women." In *Female and Male in Borneo: Contributions and Challenges to Gender Studies*, edited by Vinson H. Sutlive and G. N. Appell, 295–316. Williamsburg, VA: Borneo Research Council.

Keeler, Ward. 2017. *The Traffic in Hierarchy: Masculinity and Its Others in Buddhist Burma*. Honolulu, HI: University of Hawai'i Press.

Li, Tania M. 2015. "Social Impacts of Oil Palm in Indonesia: A Gendered Perspective from West Kalimantan." *CIFOR Occasional Paper* 124:51.

MacKinnon, J., and Kathy MacKinnon. 1986. *Review of the Protected Area System in the Indo-Malayan Realm*. Gland, Switzerland: IUCN.

MacKinnon, Kathy, Gusti Hatta, Hakimah Halim, and Arthus Mangalik. 1996. *The Ecology of Kalimantan: Indonesian Borneo*. Singapore: PeriPlus Editions.

Mashman, Valerie. 1993. "Warriors and Weavers: A Study of Gender Relations Among the Iban of Sarawak." In *Female and Male in Borneo: Contributions and Challenges to Gender studies*, edited by Vinson H. Sutlive, 231–270. Shanghai, VA: The Borneo Research Council.

Mashman, Valerie, and Patricia Nayoi. 2015. "The Bidayuh of Sarawak, Gender, Spirituality and Swiddens." In *Shifting Cultivation and Environmental Change: Indigenous People, Agriculture and Forest Conservation*, edited by Malcolm Cairns, 958–971. London: Earthscan.

McKay, Steven, and Don Eliseo Lucero-Prisno III. 2012. "Masculinities Afloat: Filipino Seafarers and the Situational Performance of Manhood." In *Men and Masculinities in Southeast Asia*, edited by Michele Ford and Lenore Lyons, 20–37. New York: Routledge.

Newton, Barbara J. 1977. "Perceptions of Sex Roles at the University of Hawaii." *Paper presented at the Women in Communication Convention*, Honolulu, HI.

Nyawalo, Mich. 2011. "Postcolonial Masculinity and Commodity Culture in Kenya." In *Global Masculinities and Manhood*, edited by Ronald L. Jackson II and Murali Balaji, 124–140. Urbana, IL: University of Illinois Press.

Ong, Aihwa, and Michael G. Peletz, eds. 1995. *Bewitching Women, Pious Men: Gender and Body Politics in Southeast Asia*. Berkeley, CA: University of California Press.

Tsing, Anna Lowenhaupt. 1990. "Gender and Performance in Meratus Dispute Settlement." In *Power and Difference: Gender in Island Southeast Asia*, edited by Jane Monnig Atkinson and Shelly Errington, 95–125. Stanford, CA: Stanford University Press.

Tsing, Anna Lowenhaupt. 1993. *In the Realm of the Diamond Queen*. Princeton, NJ: Princeton University Press.

Tsing, Anna Lowenhaupt. 1996. "Telling Violence in the Meratus Mountains." In *Headhunting and the Social Imagination in Southeast Asia*, edited by Janet Hoskins, 184–215. Stanford, CA: Stanford University Press.

Tsing, Anna Lowenhaupt. 1999. "Becoming a Tribal Elder, and Other Green Development Fantasies." In *Transforming the Indonesian Uplands: Marginality, Power and Production*, edited by Tania Li, 159–202. Amsterdam, Netherlands: Harwood Academic Publishers.

Urano, Mariko. 2010. *The Limits of Tradition: Peasants and Land Conflicts in Indonesia*. Kyoto: Center for Southeast Asian Studies, Kyoto University.

Whitten, Tony, Roehayat Emon Soeriaatmadja, and Suraya A. Afiff. 1996. *The Ecology of Java and Bali*. Singapore: Periplus Editions.

Whittier, H. L. 1973. "Social Organization and Symbol of Social Differentiation: An Ethnographic Study of the Kenyah Dayak of East Kalimantan (Borneo)." PhD, Anthropology, Michigan State University.

Zahorka, Herwig. 2017. "Punan 'Gita,' Penan Benalui, Punan Aput: From Hunter-Gatherers to Average Citizens: Early and Later Experiences of the Author in East Kalimantan." *Borneo Research Bulletin* 48:283–319.

5

MASCULINITIES IN CONFLICT IN WEST SUMATRA[1]

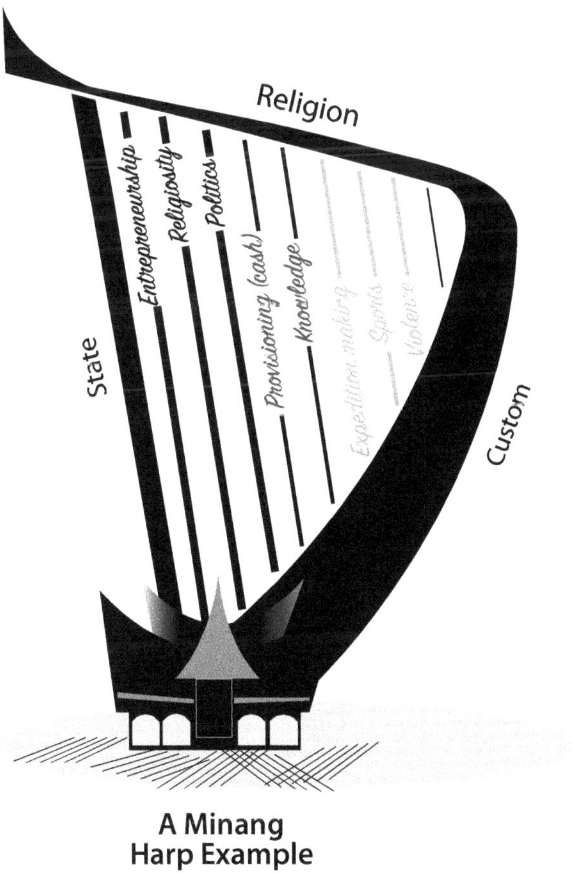

A Minang
Harp Example

Reflection and introduction

In July 1983, I moved from Hawaii's urban centre, where I'd been working as a 'Women in Development' specialist, to the rural forests of Sitiung, West Sumatra, initially as a 'Farming Systems Specialist' in an interdisciplinary team trying to improve agricultural production (Figure 5.1). Our initial, US-based team was composed of myself and two men: an agronomist, near retirement, from Island University[2] (team leader) and a soil scientist from Southern State University. These men were complemented by a team of Indonesian soil scientists from the Soils Center on Java—all men and all save one, quite junior. We were implementing a farming systems approach, which involved, among other things, close collaboration with local farmers and adaptive and holistic views of their farming systems.

My own situation differed again from previous research experiences. Now, I was part of an on-site research team, requiring me to take into account the needs and interests of colleagues. Initially, I stressed my need for daily contact with local people, both to my American and Indonesian colleagues. One of my American colleagues was amused, the other understood; both were willing to humour me, despite the inconvenience of village living to them and their families. Our Indonesian colleagues/partners had hoped we would reside several hours by car

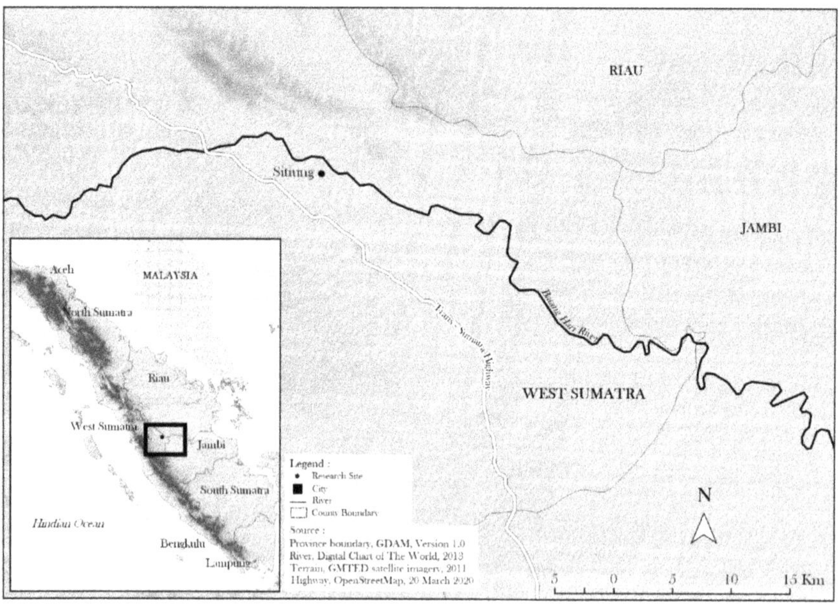

FIGURE 5.1 Map of Sitiung, within the island of Sumatra.

Source: C. J. P. Colfer, *Toward Sustainable Agriculture in the Humid Tropics: Building on the TropSoils Experience in Indonesia*, TropSoils Bulletin No. 91–02, North Carolina State University, 1991.

from our field sites, in some unused government housing; they were a bit harder to convince.

I have often experienced the paternalism of high status, Javanese men (*protection*, *caring* and *control* harp strings). They were concerned, genuinely, that we, as visitors (and me, a woman, particularly), would be uncomfortable in a village setting. We later learned how my insistence on being near 'the people' also complicated the housing arrangements for our collaborators. For a time, Indonesian team members were housed in one room of a cinder block building, sleeping on mats on the floor with no amenities whatsoever. Again and again, my status as an educated white American (with a generous research budget) intersected with and overshadowed my gender, in terms of privilege.

In the end, we rented two houses in Sitiung 1A (formally, Piruko),[3] which was the longest-established community[4] in the transmigration area: one for me and my son and one for our soil scientist with his wife and four children. Our team leader lived on the Trans-Sumatra highway, ten minutes away when the road was in good condition. Later, in 1985, I married an American man who joined me, as did my then-15-year-old daughter.

I found Minang masculinities most fascinating, because of the cultural mixture of matrilineality, matrilocality and Islam. I lived in a Javanese community adjacent to a Minang one, so had ample interaction with the men of both these ethnicities over the three years I lived there. My understanding of the Sundanese[5] masculinities found in Sitiung 5 is shallower than that of the other two ethnic groups.

In my late 30s, and by the end of the research, newly married to a monogamous man with conventional views on sexuality, I found myself far less lonely. I also had become good friends with my co-workers. But these friendships and teamwork, my private housing and my subsequent marriage also meant I spent less time than previously with the people whose lives I sought to understand. The fact that three ethnic groups (Sundanese, Javanese and Minangkabau) and five languages (those three plus English and Indonesian) were involved also interfered. Formal studies played a bigger role than previously, despite my use of participant observation here as well. My knowledge of the masculinities in this environment feels shallower than for Bushler Bay or Long Segar, though my understanding of inter-ethnic relations and their impacts on gender grew. Writing clearly about masculinities in Sitiung has proven more challenging than in the previous two contexts.

My Central Javanese neighbours had been moved en masse in 1977–1978, due to dam construction in their home area of Wonogiri. Our team initially also worked in Sitiung 5 (officially Aur Jaya) with East Javanese and Sundanese transmigrants and Minangkabau *Tran Lokal* ('local transmigrants'). 'Transmigrants' are people settled by the Indonesian government on islands off Java and Bali. In this transmigration area, transmigrants included Javanese and, in Sitiung 5, also Sundanese (both from Java). *Tran Lokal* was a programme to reduce antagonism to the transmigration programme by providing benefits to the poor in local communities (here the Minang).

These folks had all newly moved to this remote, heavily forested and only seasonally accessible community, as part of a more typical transmigration programme.[6] Gradually, as time went on, we were able to expand our collaborations to include the original inhabitants, the Minangkabau—in Pulai (adjacent to Sitiung 1) and Koto Padang (home village of the *Tran Lokal* in Sitiung 5). Residents of Pulai and Koto Padang previously owned Sitiung 1 and 5 lands, respectively.

When we arrived in Sitiung, I imagined that I understood conditions in Indonesia. I'd worked in the country for over a year and spent a fair amount of subsequent time analyzing results from Kalimantan. But Sitiung presented a very different face—one reminiscent of the Makassar in Gibson's (2005) Makassar–Buid (Philippine) comparison (see also Gibson 2011). The history of the Makassar included kings and aristocracy and the various accoutrements of a dominant 'civilization' (à la Scott 1990). Buid were an upland group that would once have been called 'primitive' and remained marginalized by the wider society. Gibson summarizes their differences thusly: "The values of social equality, individual autonomy and moral solidarity coexist in most societies with their opposites: hierarchy, dependency and factional loyalty" (p. 233). The Buid and the Kenyah tended toward the former (widely termed 'uplanders'); the Makassar, Javanese, Sundanese and to a lesser extent, the Minang, toward the latter (glossed as 'lowlanders').[7]

I make these observations to set the stage by highlighting (a) the broad cultural differences between the people of Sitiung vis-à-vis the Kenyah (Chapter 4); and (b) more centrally to this chapter, the cultural harp frames available to Sitiung's residents with differing [high] cultural histories/traditions. I have found that interactions among Indonesian ethnic groups with differing power and prestige can result in greater gender differentiation and cultural emphasis than when such interactions are absent (see also similar examples in Chapter 7 on the more recent Kenyah contexts, or Colfer et al. 2015, for such contrasts in southern Sulawesi).

Forests, soils and biodiversity

These communities, like Long Segar, were surrounded by lowland tropical rainforest. Caudle and McCants (1987) summarize the context thusly:

> Sitiung's position is 1°S latitude, 100 m elevation; it has a mean annual temperature of 26°C, and mean annual rainfall of 2,471mm, with a weak dry season, undulating topography, virgin rainforest vegetation, and soils that are predominantly clayey Ultisols and Oxisols.
>
> *(p. 121)*

There is a brief dry period, usually in February, with a more sustained dry season in July and August.

Whitten et al. (1984), writing about lowland forests of Sumatra more generally, identified the key emergents (some reaching 70 m) as Dipterocarpaceae

(*Dipterocarpus, Parashorea, Shorea* and *Dryobalanops*) and Caesalpiniaceae (including the dramatic *Koompassia, Sindora* and *Dialium*). In an area near the Batang Hari River (on which Pulai's land was located), the dominant families were Dipterocarpaceae and Olaceae, particularly *Scorodocarpus borneensis* (p. 262).

Large areas of Sitiung, including Sitiung 1, had been cleared by bulldozers beginning in 1977, resulting in removal of topsoil, soil compaction and erosion. One experiment designed to rehabilitate these soils for crop production concluded that "this soil cannot be reclaimed without chemical fertilizer and lime applications. No tillage practice will be effective unless accompanied by chemical fertilizers" (Cassel, Makarim, and Wade 1987, p. 124). Sitiung 5 farmers had the option of having their lands cleared by bulldozer—as some did—or clearing it themselves by hand (a very labourious task, primarily of men).

These highly weathered soils were acidic, many having pH of less than 4.5, with exchangeable aluminum greater than 2 meq/100 ml soil, and aluminum saturation greater than 60%. Attempts were made to address this problem by applying lime, both no-till and hoeing (Wade, Kamprath, et al. 1987), phosphorous (Wade, Santoso, et al. 1987), green manures (Wade, Heryadi, and Gill 1987) and potassium (Gill, Adiningsih, and Kasno 1987)—with green manure being one of the more effective additives.

Forest animals found in lowland Sumatran rainforests included tigers (*Panthera tigris*), elephants (*Elephas maximus*), tapirs (*Tapirus indicus*), rhinoceros (*Dicerorhinus sumatrensis*), orangutan (*Pongo pygmaeus*), siamang (*Hylobates syndactylus*), gibbon (*H. lar/agilis*), pig-tailed macaques (*Macaca nemestrina*), long-tailed macaques (*M. fascicularis*), leaf monkeys (*Presbytis melalophos* and *P. thomasi*), black giant squirrels (*Ratufa bicolor*), hornbills (*Rhinoplax vigil, Buceros rhinoceros* and *Anorrhinus galeritus*) and the great argus pheasant (*Argusianus argus*) (Whitten et al. 1984, p. 318). Although rarely seen, tigers and tapirs were definitely present locally; gibbons and siamang, common in the forest canopy, could be heard every morning in Sitiung 5 and periodically in Sitiung 1. Leaf monkeys and macaques were also evident, as were hornbills and squirrels.

Gouyon, Foresta and Levang (1993) likened the biodiversity implications of the then ubiquitous rubber gardens (their 'jungle rubber') that characterized local Minang farming systems to those of secondary forest.

> An assessment of biodiversity on the Jambi plot [Muarabuat, Kec. Rantau Pandan, Kab. Muaro Tebo, near Sitiung] revealed 268 plant species other than rubber, all originating from natural forest, distributed into 91 tree, 27 shrub, 97 vine, 23 herbaceous, 28 epiphytic and 2 parasitic species . . . This is equivalent to the plant diversity in an old secondary forest. A comparison with weeded, estate-like plantations that included only a few species other than rubber underlines the importance of jungle rubber for the conservation of forest plant diversity.
>
> In short, a jungle rubber plantation presents the features of a rubber-based secondary forest that usually lasts up to 40 years or more before being

replanted, while secondary regrowth seldom exceeds 20 years in a shifting cultivation cycle.

(p. 188; see also Michon, Mary, and Bompard 1986 for similar Minang agroforestry systems at higher elevations to the west)

Unlike the cases in Chapters 3 and 4, formal forest managers were not in evidence in Sitiung. Kerinci Seblat National Park, a considerable distance on difficult roads, was the closest protected area, with minimal management. I knew of no formal timber concessions nearby, though local communities did some informal (and thus illegal) logging in the area (discussed briefly later in the chapter). One small, experimental rubber plantation operated nearby, trying to convince locals to plant improved rubber in orderly rows and improve their collection methods, thereby producing higher quality rubber, and local people managed trees with edible produce as well (discussed later).

Narratives and practice of gender relations

I emphasize three chords, on many of Sitiung's songs of masculinity: The first loosely links the religion, sexuality and parenthood harp strings. The second considers politics and hierarchy. The third examines breadwinning and provisioning. A fourth section examines other individual harp strings with less centrality: expeditions, knowledge, violence, sports and arts.

Different men can create different chords from among the harp's strings, with varying ethnic tendencies/preferences. Clearly the interconnections within these examples of possible chords vary in their intensity, from person to person and from time to time, and there can be overlapping effects as one plucks one string or another, one chord or another. Such interactions among elements characterize any system (including gender systems). Figure 5.0 reflects a common Minang song. The discussion in this section recounts the kinds of options visible to my 'outsider' eyes.

In each section, I differentiate the Minang and transmigrant masculinities, occasionally differentiating Javanese and Sundanese patterns within transmigrants. Ethnicity here plays a role comparable to occupation in Chapter 3, as the most significant social structural differentiation. At the end of this chapter, like the others, I summarize the dominant harp strings and the links with forests.

Sitiung was characterized by serious resource competition and conflict between the Minang and the transmigrants, whether Javanese or Sundanese.[8] The transmigrants came as part of the government's transmigration programme designed partly to 'civilize' Indonesia's 'Outer Islands' (à la Geertz) and earlier, to reduce Java's population pressure. They had governmental support and were part of a coherent and long-lasting national policy. The Minang were the long-term inhabitants of the region, losing out to rather unwelcome in-migrants, foisted off on them by the central government and distant Minang decision-makers, mainly from the Minang heartland.

Although physical violence was rare among ethnic groups in Sitiung, there was significant mutual antagonism and adverse stereotyping.[9] Javanese tended to see Minang as lazy, sneaky and dishonest, with very peculiar social structure and extreme religiosity; Minang saw Javanese as unwelcome invaders, immodest and bad Muslims, with a peculiar affinity for hoeing and physical labour. Kahin (1999) recounts a famous Minang legend unsuccessfully pitting the Javanese against the Minang, leading to a Minang "conviction of superior intelligence to the Javanese" (p. 24). Elmhirst (2018) provides a nice summary of the Javanese orientation, which assumes Javanese superiority. Such views were also operative in 1980s Sitiung. Differing ideas and practices related to masculinity contributed to these stereotypes, as should become clear next.

Religiosity and parenthood harp strings and gender dynamics

Initially, the most surprising element of Minang masculinities was its overt religiosity. Knowing that the group was matrilineal,[10] I had not realized how important Islam was to them. With my Middle Eastern background, and its more conventional, strongly patrilineal version of Islam, I found the combination of matrilineality and Islam jarring. How could the two be combined?[11]

Islam's role in Minang daily life had penetrated more deeply than had Christianity among the Kenyah of East Kalimantan. The Kenyah had been given three choices in 1965, when Indonesian soldiers had come to central Borneo as part of the confrontation with Malaysia: accept Islam, Christianity or die. Christianity allowed continued consumption of pig, central to their cuisine, so they chose it. Islam, on the other hand, had been dominant in West Sumatra for centuries. And with that had come a more elaborate ideology of women's inferiority and a stronger valuation on manhood.

A central conflict between the Minang and the in-migrants pertained to religion, and much of that was expressed as disapproval of the other's gendered behaviour and ideas. The matrilineality of the Minang was considered peculiar and vaguely un-Islamic by the transmigrants, but stronger disapproval came from the Minang, who saw the Javanese lack of concern about women's (and to a lesser extent, men's) modesty as evidence of gross disregard of Islamic strictures.

A few Javanese men did pluck the religious harp string. But few Javanese community members observed Islamic rules very seriously. Many willingly ate pork when it was available, and neither men nor women were particularly modest about their clothing; a few older men wore the black religious hat. Community members tended to link overt religiosity with the educated, who admired its public expression. My Indonesian soil scientist co-workers (of mixed ethnicity, virtually all educated men), for instance, attended the mosque every week, fasted during Ramadan and some prayed regularly. They were proud of being good Muslims.[12]

Such religious observances represented a crucial ethnic differentiation for community members however, with the Minang men considering themselves far 'better Muslims' than the transmigrants. Besides Minang observance of Islamic strictures (regular mosque attendance, fasting, praying routinely and observing Islamic dietary restrictions), the Minang objected to what they considered Javanese immodesty, particularly among women. Nudity was simply not a big deal among the Javanese. One of our neighbours, an old man, liked to bathe nude by our outdoor well, holding a hand over his genitals; his wife and younger neighbour women wandered around the neighbourhood with only a bra on their upper torsos. Minang men particularly found such practices offensive and counter to Islam.

But within their own communities, Peletz's (1995) observation on the Malays of Negeri Sembilan—"In most social and cultural contexts gender is of relatively little concern and does not constitute a highly salient marker of social activities or cultural knowledge" (p. 81)—applied to all these ethnic groups as well. In 1985, we conducted a cognitive mapping study ('Galileo') similar to that reported in Chapter 4,[13] confirming this observation. We found the concepts, *man* and *woman* vis-à-vis *good* to be very similar within each ethnic group, with *woman* consistently slightly closer to *good* than *man:* for the Minang, *man* averaged 49 units from *good*, *woman*, 46;[14] for the Javanese, *man* averaged 25 units from *good*, *woman*, 24; and for the Sundanese, *man* was 27 units from *good*, *woman*, 26 (see Tables 5.1 and 5.2). In the nine open-ended interviews conducted to identify the relevant concepts, only one interviewee mentioned any gender differentiation.

TABLE 5.1 Measure of community values and soil management concepts by ethnicity, Sitiung, 1985

Good and Fertilizer	*Minang*	*Javanese*	*Sundanese*
Soil	s 60	m 31	m 23
Garden	s 60	m 27	m 19
Unirrigated field	s 59	m 27	m 25
Wet rice field	s 58	m 30	m 23
Home garden	s 56	m 28	m 20
Rubber	s 64	ms 41	mj 26
Fruits	s 56	m 29	m 20
Rice	s 55	m 25	m 24
Other field crop	s 62	ms 30	mj 19
Vegetables	s 62	m 24	m 23
Water	s 46	m 25	m 24
Fertilizer	s 34	s 28	mj 16
Pests	s 68	ms 52	mj 26
Yield	s 61	m 26	m 27
Cultivation	s 66	m 23	m 24

Note: The smaller the number, the closer the concepts. If there is a letter before the number, there is a significant difference between ethnic groups. In the Minang column, the distance between "good" and "soil' (60) is significantly greater than that perceived by the Javanese (j) and the Sundanese (s).

Source: Colfer, Newton, and Herman (1989); C. J. P. Colfer, *Toward Sustainable Agriculture in the Humid Tropics: Building on the TropSoils Experience in Indonesia*, TropSoils Bulletin No. 91–02, North Carolina State University, 1991.

TABLE 5.2 Measure of community perceptions of men and soil management concepts by ethnicity, Sitiung, 1985

Men and	*Minang*	*Javanese*	*Sundanese*
Garden	14	22	19
Unirrigated field	17	22	26
Wet rice field	13	23	26
Home garden	26	24	22
Rubber	js 14	ms 50	mj 32
Fruits	29	24	23
Rice	j 10	m 24	16
Other field crop	33	24	23
Vegetables	s 34	s 28	mj 16
Water	s 52	s 56	mj 21
Fertilizer	js 47	m 27	m 25
Pests	s 50	s 55	mj 28
Yield	js 44	m 26	m 25
Cultivation	23	16	18

Note: The smaller the number, the closer the concepts. If there is a letter before the number, there is a significant difference between ethnic groups. In the Minang column, the distance between "men" and "rubber' (14) is significantly smaller than that perceived by the Javanese (j) and the Sundanese (s).

Source: Colfer, Newton, and Herman (1989); C. J. P. Colfer, *Toward Sustainable Agriculture in the Humid Tropics: Building on the TropSoils Experience in Indonesia*, TropSoils Bulletin No. 91–02, North Carolina State University, 1991.

However, Peletz goes on to say "there are certain areas in which gender differences are culturally elaborated" (p. 82), such as that between reason (*akal*) and passion (*nafsu*). Within Islam, *akal* is linked to men, *nafsu* to women (also noted by Laderman 1996). However, as Brenner (1995) has shown regarding Javanese men, these qualities are somewhat reversed in everyday life—a pattern visible in all three Sitiung communities. Spiller (2010) reports the Sundanese perception of 'misbehaving men' as 'cute' squirrels (p. 34, unlike their common American representation as 'rats'; see also Lentz 2017; or Hall 2011, on 'naughty boy' expectations for men in Jamaica; Barker et al. 2011, similarly for men in Brazil).

All groups tended to link the idealized men's *akal* with spiritual strength, not necessarily related to Islam. More conventionally magical beliefs were engrained along with Islam. My Minang field assistant, Syarif, for instance, linked love with *ilmu sihar*, black magic:[15]

> [I]f someone fell in love with [a woman intern on our project] and . . . [she] refused him, he might use magic on her to make her *bodoh* [stupid].[16] . . . as with the woman he'd told me about who'd been made afraid of people and unwilling or unable to talk.
>
> *(notes, 11 June 1986)*

Notions of *akal* and *nafsu* link directly to sexuality (see Krier 1995, on Minang views of young men's sexuality as dangerous and threatening, with *akal* bringing it under control) and gender dynamics. In one case, Pulai community members used religious grounds to blame an unmarried mother's parents for her condition, as well as the girl herself (notes, 1 May 1986). Among the Minang, neither boys nor girls under 15 were seen to sin; they were not yet held 'responsible' (notes, 8 January 1986).

Fatherhood was important to men of all three ethnicities, though the emphasis tended to be more on parenting by the couple as just described. A group of Minang men asked me once why my husband had married me if I could no longer have children.

> They said Minang men wouldn't want to marry a woman with whom they couldn't have children. Taking care of another man's children didn't appeal to them.
>
> *(notes, 8 January 1986)*

The importance of both boys and girls was reflected in a Minang proverb, with reference to having only one child: "He/she[17] is my son and my daughter", which they further explained: "There is only one narrow thread and when it breaks, that's it" (notes, 11 June 1986). The lineal nature of the Minang kinship system likely strengthened the importance of having at least one child of one's own. The Javanese and Sundanese, both bilateral in their kinship, readily adopted, and though men and women generally very much wanted and delighted in children, I did not sense the same commitment to a sort of genetic or familial continuity.

Physical strength is an important masculine feature among many groups. The Minang recognized men's greater strength, likening it to that of male vs. female water buffalo (notes, 18 June 1986). Such strength was not, however, an overt matter of sexual attraction. As with the Kenyah, Sitiung girls and women (and men, for that matter) did not show admiration for men's muscles (unlike in the US).

Among the Minang, the process of choosing a spouse was said to begin with either the parents of the girl or the boy; then the opinion of the young people was sought (notes, 22 November 1985).[18] However, among all three ethnic groups, there was ample opportunity for young people to meet (though not to 'date' in the American fashion) and form opinions about each other's desirability. Minang agricultural work parties, for instance, could provide opportunities for young people to dance and sing together late into the night. The Minang did not allow marriage within one's clan, but young people's wishes were usually considered seriously in marriage decisions in all three groups.[19]

A Minang truck owner/driver in Sitiung 5 talked about his views on the marital relationship (presaging the breadwinning discussion later). He considered it

> sensible for a man to be mad at his wife if he came home and there was no rice cooked. He felt in the long run though that either, if wrong, should

> just be quiet when the other was angry. The wrongdoer should temporarily leave the house and return later. If the wrongdoer argued, there'd be a divorce, and that wasn't good for the kids, who'd grow up with no source of money for schooling and whatnot, since men are usually the obtainers of money. . . . He told me I shouldn't argue with my husband.
>
> *(notes, 29 January 1984)*

One of my American soil scientist colleagues asked a group of Minang men testing soils together if women were ever smarter than men, something they denied. He then suggested perhaps women's use was being pretty, to which they responded, "What's the good of pretty if they're not hard working?!"—rather like Kenyah women's views of men (and of each other). Minang women spent a *lot* of time on agricultural labour; Javanese and Sundanese men (and women) conversely highly valued and stressed *men's* hard work, especially with regard to agriculture and linked it with both men's physical strength and their gender role. These different patterns were a further source of mutual ethnic antagonism and disrespect.

Data from a time allocation study[20] provided evidence that men's control of women among the Minang, despite their being a matrilineal group, may have been greater than among transmigrants coming from Java or the Kenyah:

> We have postulated elsewhere (Colfer et al. 1984) that the amount of leisure activity, by sex, might be one index of gender equality within a given culture. In Pulai, this ratio (186 observations of females at leisure to 237 of males at leisure) computes to .78. The comparable ratio for transmigrants was .89; and for a very egalitarian group in Kalimantan (the Kenyah; see Chapter 4), .98.
>
> *(Colfer 1981, p. 44)*

The Minang men's strategy of waiting for economic opportunities to come along the road (discussed in the section on breadwinning) means that their public 'leisure' has the potential to become 'productive', however.

Political and hierarchy harp strings

The Sitiung folk all have histories of 'high' culture, ones in which the *hierarchy* harp string was key along with many other features identified with 'lowland groups' by Scott (2009) and contrasting with 'uplanders' like the Kenyah. The Javanese and Sundanese transmigrants come from long lines of powerful kingdoms, going back centuries. The 15th-century Minang kingdom of Pagarruyung was more recent, centrally located in the Minangkabau heartland, far from Sitiung and more a loose federation than a 'proper' kingdom. Kahin (1999) notes that "in practice, [the 'king'] had no authority or executive, legislative, or juridical power whatsoever with respect to the *luhak* [districts] and *nagari* [described below]" (p. 22)—suggesting a more *egalitarian* harp. She contrasts the Javanese and Minang approaches:

> The traditional Javanese belief is that the well-being of the state depends on the strength of the ruler at the centre [á la Ben Anderson] . . . In the traditional Minangkabau view, in contrast, the welfare of the polity depends on the harmony and agreement among its components—the extended villages (*nagari*) . . . These *nagari*, which have often been described as 'village republics', exercise their own political and consensus democracy, a form of government very distant from that of the Javanese ideal.
>
> *(p. 16)*

Kato (1982) describes the *nagari* as "the highest order of human settlement acknowledged by the *adat* [custom]" (pp. 41–42). The local Minang system contained multiple hierarchies that differed from those of the transmigrants. They were also considerably more obvious and more formally interconnected than those of the Kenyah—via longstanding adherence to Islam (a world religion), governance structures spanning various and intersecting scales and formalized customs widely shared.

I begin here with the Minang political realm because this is the arena in which forests co-evolved in this region. However, the more agriculturally oriented Javanese system is closer to what the central government was attempting to spread throughout the country in the 1980s.

Minang political and hierarchy harp strings

Politics was an important harp string for local Minang men. Before agreeing to my conducting research in Pulai, the senior men grilled me about whether I'd be engaging in *politik*, by which they meant primarily attempts to convert them to Christianity. The questions they asked led me to conclude that they were "rather political, savvy sorts of men" (notes, 15 October 1985).[21] There was a backstory to this suspicion, an event that began right before I approached the community to conduct research there.

BOX 5.1 RELIGIOUS CONFLICT IN WEST SUMATRA

In mid-1985, one of our team members kindly (and innocently) picked up a couple of families of Christians in Sitiung 5 to take them to church in Sitiung 1. Only later did we learn that Christians were not allowed to join this transmigration project (a stipulation of the provincial decision-makers), though a few Christians had snuck in among the transmigrants. This became an issue shortly thereafter when a Minang man came to my colleague's home in Sitiung 1, interrupting a housewarming party for the members of this tiny Christian 'church'. The Minang visitor wrote a report to the *Bupati* full of complaints about our project, some—like providing electricity from our generator to our neighbours—quite nonsensical. We went to Sawahlunto-Sijunjung, the district capital, to explain. The letter went up to the governor. We went to Padang to explain. It then went to Java, where it made its way eventually to the American embassy. Despite our serious efforts to resolve the issue at all levels our colleague was removed to Java for six months and then expelled from the country—among the more unjust events I've witnessed in my life.

Many of Pulai's men were more intimately integrated into the national political hierarchy than women, the Kenyah or the men of Bushler Bay. This integration was reflected in their insistence on confirming repeatedly that I had permission from the *Camat* (county head, part of Indonesia's national administrative structure) to conduct the research.

Such integration was especially intriguing because national governmental policies did not mesh well with either matrilineality or matrilocality.[22] Indonesian national policy recognized men as heads of their marital households (Elmhirst 2011; see also Jakimow 2017), not their sisters' households, nor did it recognize clans or *kenagarian* (the traditional Minang supra-village organization).[23] Minang men have been unusually astute at manoeuvring national politics, due partly elsewhere to higher than average educational achievements and to their engagement in expedition-making, where they gained linguistic skills and established broader-scale relationships (see Kato 1982).

At a graduation from Pulai's small religious high school just prior to the 1986 national election, virtually all the officials, from the distant *Kabupaten* (district) capital attended, as well as the local *Camat* and the heads of three of the four Minang clans[24] in that village (notes, 24 April 1986)—all these officials men, all or almost all Minang themselves.[25] Only young women served the assembled multitudes;[26] and a beautiful young girl had been chosen, attended and dressed as their *Bundo Kanduang*, the key symbol of Minang womanhood, along with an older, also beautifully dressed woman I took to be the previous *Bundo Kanduang*.

A speech given by the *Bupati* (head of the Regency) nicely reflected important Minang masculine ideals (education, breadwinning, striving, religion, family):

> First, he told how you had to keep learning your whole life, how you mustn't be unemployed even if you couldn't continue school, how those who hadn't passed shouldn't feel '*kecil hati*' ['small liver', demoralized] but rather just try harder, that this was a test from God. They should work and do their duties to their communities and try to help their parents.
>
> *(notes, 24 April 1986)*

That a man's status depended on his own accomplishments as much as social hierarchy was also clear from his speech—reminiscent of American success stories:

> He himself [the *Bupati*] hadn't had enough to buy one shirt [in earlier times], he said, but now he does. He [said he] makes Rp. 250,000 per month in salary and the newest employee can now make Rp. 40,000.
>
> *(notes, 24 April 1986)*

Hierarchy was hardly however irrelevant, as shown by the repeated requests, during early days that I confirm permission from clan leaders, the village head and the *Camat*.

Pulai's four matrilineal clans were ranked (based on arrival in the area),[27] and the men within them (the brothers of the land-owning women) managed clan affairs and governed the communities politically. The permission of the leader of the first clan, Tigo Ninik, was needed before clearing new lands. Traditionally, Pulai and three other villages were *jorong* (hamlets) of one *kenagarian*[28] (previously called Sitiung), a broader Minang political unit (see Firmansyah 2013; Kato 1982 or Naim and Hermann 1984, for descriptions of this hierarchical system and its variations).

Transmigrant hierarchy and equity harp strings

The very elaborate public governmental hierarchy among the incoming transmigrants (and others) was manned by men. Sitiung 1's four neighbourhoods (*jorong*, ca. 100 households each) comprised a village (*desa*), within an area led by a *wali nagari*, smaller than a *county* (*kecamatan*, led by a *camat*), within a district (*kabupaten*, led by a *bupati*), within a province (*propinsi*, led by a *gubernor*) (notes, 17 Sept 1983). All such officials were men. As a *bedol desa*[29] transmigration site, the original economic inequities of the Central Javan community from Wonogiri were replicated in Sitiung 1. Landowners (mainly men) were compensated based on existing holdings.

Sitiung 5 was a more conventional transmigration site, composed of resource-poor participants from East Java who had applied to join. The Sundanese transmigrants were more diverse in terms of wealth, having transmigrated after Galunggung's volcanic eruption in 1982 from Garut Province in West Java. In Sitiung 5, household heads were all men.

The assumption and approval of hierarchy spread throughout Javanese social interaction. Our collaborating Javanese (and Sundanese) soil scientists initially complained when the American team members worked in the field with the farmers. The former saw such work as incompatible with their educated and elite statuses (notes, 19 Oct '83)—a view also expressed by Turks during my childhood in Turkey. Many Americans, whether educated or not, found hard physical labour totally compatible with their own notions of masculinity (see Chapters 2 and 3).

Javanese—both men and women—often reiterated the necessity of following governmental dictates. The views that Sumartoyo from Sitiung 2 (another nearby transmigration area) expressed about submission to authority were common (so incompatible with the Bushler Bay Local view of masculinity). When asked if he preferred living in Sitiung 2 or his home community in Wonogiri, Central Java, he chose the latter.[30] But he added,

> The government ordered me to move over here, and well, I had to come. If the government hadn't ordered me, I really didn't want to move here. I am a citizen; how could I refuse the government policy?
>
> *(Constraints Study notes, 1985)*[31]

This attachment to hierarchy and willingness to submit were moderated by a concern for equity. There was a lot of complaining among Javanese men about the immorality (infidelity, lack of industriousness, dishonesty) of others and about the unfairness of events or distributions. Some of this could be attributed to understandable attempts to discredit others in order to obtain jobs or benefits themselves, but some seemed to be genuine moral pronouncements on the behaviour of others, as Elmhirst (2018) also found in southern Sumatra, among the Lampungese.[32]

In my American experience, gossip was 'not manly'; it was considered feminine, petty and somewhat improper, 'poor form'. But Javanese men were quite willing to so engage. Complaints about inequity were also avoided by American men as evidence that one had not been in control, had not been able to get a fair deal for oneself, in itself an affront to one's masculinity.

Some of the more lucrative breadwinning activities depended upon good political connections. The political activities of Sitiung's men, especially their broader-scale involvements and their commitments to varying hierarchies, represented another important difference from Bushler Bay and Long Segar.

Breadwinning and provisioning harp strings

Although both Minang and transmigrant men saw breadwinning as part of men's roles and values, how this played out differed. Here, I begin discussing the Minang system, shifting in the second half to the transmigrant systems.

Minang breadwinning/provisioning

There were certain similarities between Minang and Kenyah divisions of labour, in that Minang women were also actively involved in rice production and Minang men were also clearly expected to provide cash. Still, Pulai's men also engaged in a significant amount of agricultural labour (see Figure 5.2). In addition, they were actively involved in forest harvesting, fishing, hunting and wage labour—all dependent on the surrounding forest. All in all, Minang men were twice as involved in 'productive activity' as women, reinforcing the notion that being a breadwinner was an important element in local masculinities.

Although dependent on agriculture, Minang men tended neither to consider it a desirable breadwinning strategy nor to identify being a farmer in a positive light. In Tables 5.1 and 5.2, Minang respondents considered these agricultural concepts

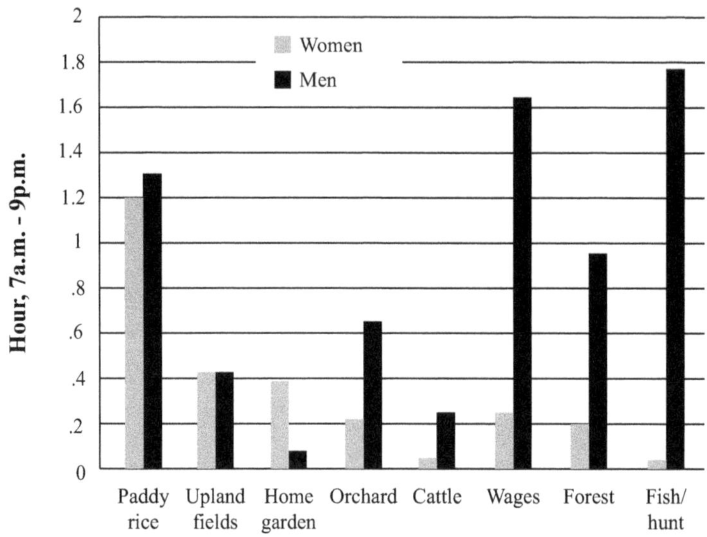

FIGURE 5.2 Time allocation to eight productive activities by sex, Pulai, 1985–1986.

Source: C. J. P. Colfer, *Toward Sustainable Agriculture in the Humid Tropics: Building on the TropSoils Experience in Indonesia*, TropSoils Bulletin No. 91–02, North Carolina State University, 1991.

further apart on average than did the more agriculturally oriented Javanese and Sundanese. Arni, a Minang woman in Sitiung 5, expressed her view: "Let's *angkat kayu* [engage in timber transport] if there's a chance, and if we can't do that, well, just work in the *ladang* [upland rice field]" (notes, 5 December 1983).

Whereas in the Minang heartland, land and houses normally belonged to women and their clan (*suku*),[33] in Sitiung, people described a two-part division: (1) paddy rice lands (*pusoko*), belonging to women, managed by brothers (who inherited clan titles, *soko*), and inherited by women's daughters; and (2) rubber and other dry lands belonging to the man who cleared the forest and which could be passed down to his own children.[34] Matrilocality was the rationale given for women owning land (notes, 3 December 1985). However, when we conducted a survey of land ownership in Pulai, in June 1986, people reported most land to be owned jointly, except for rubber gardens (over half of which were reported as owned by women!). See Suyanto, Tomich, and Otsuka (1997) for a nearby and somewhat similarly complex, fluid and [to outsiders] confusing system. The distribution of income from rubber gardens is also of interest, with men owning fewer fields and hectares, but gaining more income (see Tables 5.3 and 5.4).

Turning to the domestic context, Minang men were somewhat marginalized emotionally.[35] One visit to a Minang household sticks in my mind. An old mother sat at home with several of her middle-aged daughters chatting. A

TABLE 5.3 Ownership and hectarage of Pulai fields, using local field categories, June 1986

	Total		*Women's*		*Men's*		*Joint*	
	No. of fields	*Total ha*	*No. of fields*	*Total ha*	*No. of fields*	*Total ha*	*No. of fields*	*Total ha*
Paddy rice field (*sawah*)	63	18.9	11	3.2	2	0.5	50	15.1
Upland rice field (*ladang* and *soso'*)	28	22	0	0	2	1.5	26	20.5
Rubber orchard (*kebun karet*)	52	34.9	32	18.7	9	7.4	11	8.8
Other orchard (*kebun*)	8	2.3	8	2.3	0	0	0	0
Home garden (*pekarangan*)	82	15.1	17	3.2	4	0.4	61	11.4
Totals	233	93.2	68	27.4	17	9.8	148	55.8

Source: C. J. P. Colfer, *Toward Sustainable Agriculture in the Humid Tropics: Building on the TropSoils Experience in Indonesia*, TropSoils Bulletin No. 91–02, North Carolina State University, 1991.

TABLE 5.4 Ownership and control of income from the 52 rubber orchards reported in Pulai, June 1985 through May 1986

Owner/ beneficiary	*Ownership*		*Control*	
	Rubber trees	*Hectares*	*Income in US $*[a]	*% total income*[b]
Women (32 fields)	7800	18.7	1492	2.8%
Men (9 fields)	3100	7.4	4450	8.6%
Joint (11 fields)	3650	8.8	1422	2.7%
Total	14550	34.9	7364	13.8%[c]

Note: Thirty-seven families (43% of all Pulai families) owned these fields.

a US $1 = Rp.1127 June 1986
b Percent of total village cash income, deriving from rubber, controlled by each category (women, men and jointly)
c Rounding error

Source: Colfer, Gill, and Agus (1988); C. J. P. Colfer, *Toward Sustainable Agriculture in the Humid Tropics: Building on the TropSoils Experience in Indonesia*. TropSoils Bulletin No. 91–02, North Carolina State University, 1991.

few men wandered in. They sat against the wall, separate, back from the circle of related women. The women were laughing and talking, having fun. The men's contributions were ignored, trivialized, even laughed at. On another occasion,

> the women were working closely together, had close relationships. The men came in and out, without being terribly involved. But there was also an appreciation of them [the men] to some extent.
>
> *(notes, 19 May 1986; also noted by Tanner 1974)*

Siskind (1973) notes the marginal and ambiguous position of men among the matrilineal and matrilocal Sharanahua of the eastern Peruvian Amazon.[36]

Minang men expressed confusion as to whether Sitiung 5 land allocated to them then belonged to the men or to their wives, with one of the more powerful men considering it belonging to the men; but others saying it was their wives' (notes, 8 Nov 1983). Some of this confusion related to the disjuncture between their matrilineal system and the national identification of men (husbands, not brothers) as household heads and landowners; some to the inherent fuzziness of this system on the border between matrilineal and patrilineal kin systems.

Breadwinning for men was an entrepreneurial endeavour. In interviews on income,

> a group of three men [mid-20s to mid-40s] emphasized the opportunistic nature of their income getting activities. That nothing is set, that they're always on the lookout, always comparing the relative profitability of different endeavours, and switching their activities accordingly. . . . Min emphasized the importance of using your brain; he 'lowed as how it might look as though Minang men were just sitting around, but they weren't really. They were figuring and fenagling and working on making money. He even mentioned IQ!
>
> *(notes, 18 April 1986)*

Minang men were not known nationally for 'hard work', rather they were known for (and considered themselves to be) 'working smart'. One could regularly see groups of men just sitting and chatting by Pulai's main road (at the *pos ronda*, a small roadside structure with only a roof), apparently relaxing. However, they were also waiting for job opportunities that regularly appeared along roads—for logging, dredging for sand, driving trucks, moving goods, clearing land, etc.

Syarif, my locally admired field assistant, was jokingly called 'Ahmat Lungga' ('Ahmat Loose'). This name referred to his 'moving [*goyang*] and shaking' approach to life: "He's always doing a bit of this and a bit of that, accepting this work and that. A bit of everything" (notes, 6 February 1986). Minang men's breadwinning harp string has a strong entrepreneurial element.

Men also cleared the big trees from forest land intended for rubber and other tree crops;[37] men were almost never seen weeding or harvesting rice crops (though they *were* involved in transporting them).[38] A mixed group of people preparing a field told me though that "men and women don't have very separate work. There isn't anything, they said, that men can't do or women can't do. . . . Husbands and wives work together" (20 October 1985). On the same day, men were dibbling a swidden field and women planting rice in the field intended eventually for tree crops. Men tapped most rubber. Men did a considerable amount of fishing (with nets they made) as well, in the beautiful Batang Hari River (Box 5.2). Old men and women both ran small stores (*warung*). As is common in Indonesia, most men turned over any income to their wives to manage.

BOX 5.2 BATHING IN THE BATANG HARI

[Two men and I] went on down along the bank of the [Batang Hari] river [about 5 PM]. The river was alive with people. Everyone was bathing, talking, fishing, etc. It was lovely! The greenery came down to the water's edge in places. The weather was cool. The sounds of monkeys in the bamboo and trees along the shore reached us regularly, along with other sounds. We saw two or three fruit bats pass overhead. Everyone was interested that I was in the boat and smiled friendlily. It was so reminiscent of *sore* [late afternoon] in Kalimantan, although there are no *jambans* [bathing rafts]. The river bottom is rock—smooth rock but very hard on my feet (they found this interesting, since it doesn't seem to hurt their feet at all) . . . I loved it. Am so glad I went. The feel of doing something is so different from reading that at 5 PM people usually go bathe. Not only do they bathe, but it is a totally refreshing, comfortable, desirable feeling to look forward to all day! (notes, 18 June 1986)

Young strong Minang men in Pulai also logged, using a chainsaw and dragging the logs out of the forest by water buffalo, contracted by a Pulai leader (*lipati*). They differentiated wood for export and for local consumption and appreciated at least three kinds of *meranti*, or Philippine mahogany (Dipterocarpaceae): *meranti kawang*, *meranti ambai* and *meranti perang*, in decreasing order of value. Men also made *balok* (beams), harvested rattan in the woods and tapped rubber in their 'jungle rubber'. Work in the woods was not, however, appreciated more than other modes of making a living; indeed, less so.

Rais, a 37-year-old Minang farming man (from Koto Padang),[39] said,

> In other areas, the women can get additional money for shopping, and the men can work hard on the land. In Koto Padang, men and women are the same, they work too little, are lazy, and the soil isn't good.
>
> *(4 April 1985)*[40]

Some of the jobs Sitiung 5's Minang men did included rubber tapping, transport to market, casual labour at a nearby smallholder rubber project (notes, 4 April 1985), village rubber merchant and making furniture (notes, 8 November 1983). Although the Minang expressed a preference for non-agricultural endeavours, much of their time, both men's and women's, was devoted to farm work. There were few other opportunities.

Transmigrant breadwinning/provisioning

Among most transmigrant men—whether Javanese or Sundanese—being a farmer was a central part of their identity. But many had also to work for wages sometimes, to make money and perform another strong normative role as provider—the reverse of local Minang breadwinning prioritization. Javanese men consistently expressed interest in learning more about agriculture from our project team. A carpenter-farmer in Sitiung 1 told me, for instance, "Fundamentally, I am a farmer, but I do wage labour too. The income from wage labour doesn't last long, it's not really secure and lasting" (notes, 7 September 1983). When he was doing wage labour, his wife took up the agricultural slack. He considered these two ways to make money as alternatives in ensuring the family's subsistence—which was far more dependent on the availability of money than was the case in Long Segar (Chapter 4).

We studied annual income sources for 1983–1984 (Colfer 1991, p. 21) among 20 randomly selected families each in Sitiung 1 and 5. In Sitiung 1, 22% of household income came from home industry (probably predominantly women's work), 42% from wage labour (mainly men's) and 36% from agricultural production (both). In Sitiung 5, among a group our project was *not* working with (our 'control'), 28% of income came from home industry, 86% from wage labour and 10% from agricultural production. Our collaborating farmers in Sitiung 5 did considerably better agriculturally (possibly reflecting the selection process as much as the success of our efforts). Their income sources were 11% from home industry, 43% from wage labour and 46% from agricultural production.

Some of the unskilled off-farm jobs Sitiung 1's Javanese men did included cutting grass along the irrigation canal or digging on road or irrigation projects. Some professions in longer-established Sitiung 1 included being a tailor, traditional healer (*dukun*), driver or worker for the Irrigation Department. In newly settled Sitiung 5, men cleared land, logged and moved large logs off fields, made beams and repaired houses. There were few off-farm options in Sitiung 5, because of the remoteness of the locale and the recurring impassability of their road. Javanese women were involved in small-scale marketing, as elsewhere (Dewey 1962; Brenner 1995), though with remarkably little to sell.

Of the three ethnic groups observed in Sitiung, the Javanese had the least extreme gender division of labour (though more marked than among the Kenyah). Although there were tasks men were more likely to do than women and vice versa, there was willingness on both parts to do the work typically assigned to the other as needed. In the Javanese home, where I stayed periodically in Sitiung 5, for instance, the husband, Ratno, routinely cooked while his wife held their new baby. Both men and women were involved in both communities in making money and in agricultural tasks. Fathers and others were involved in births.

One differentiation ripe with ethnic symbolism—and promoting inter-ethnic disapproval—was the Javanese masculine affinity for hoeing (*cangkul*) and fertilizer use. The Javanese belittled the local Minang for their unwillingness to hoe. Javanese men linked hoeing with their own commitment to industriousness. The Minang, on the other hand, disdained both hoeing and fertilizer use (see Tables 5.1 and 5.2).

> I had some interesting discussions with Ratno about their land and land preparation. [T]he thought of *not* hoeing does not seem to have occurred to him.
>
> *(notes, 23 January 1984)*

> [Suminah, a Javanese woman] would want a man to *cangkul*.
>
> *(notes, 29 January 1984)*

> [Hoeing] seems to be the symbolic thing that women shouldn't have to do, and . . . Minang, Javanese and Sundanese . . . feel sorry for women who have to do that. Most [ethnic groups] maintain that other [ethnic groups'] wives have to [hoe], though I've seen [women hoeing] in all groups.
>
> *(notes, 29 January 1984)*

The fact that Minang women more often hoed—in paddy rice cultivation, where they were particularly involved—was also seriously disapproved and considered further evidence in Javanese eyes that Minang values had gone awry. Numerous examples of negative comments about Minang men's laziness were recorded in my fieldnotes (notes, e.g., 4 October, 19 November 1983). Minang men (like the Kenyah men of Long Segar) readily acknowledged this trait in themselves as well: 'working smart' rather than 'working hard', and this did *not* include hoeing on upland fields, as we discovered when we designed collaborative experiments that required hoeing a section of their fields.

Part of our Sitiung 5 work involved collaborating with farming families about their upland agricultural fields. We made great efforts to involve the farm women in the planning and report-back meetings . . . but none came. Such meetings, focused on upland fields, were in their eyes, men's work. At planting time, women came to the fields to supply labour. Indeed, they were quite involved in planting, as well as other agricultural tasks. Figures 5.3 to 5.7, based on the time allocation study described earlier, show the involvement of both genders in agriculture in all three ethnic groups. Figures 5.3 and 5.6 show the dominance of men in field crops—the context most likely to be defined as 'agricultural' and where most agricultural research was focused.

Land clearing was a job reserved for men. In Sitiung 1, there were spikes in January and August, in preparation for dry periods (when fields could be burned) and in Sitiung 5, men were busy in their upland fields, again making use of the shorter of the two dry seasons.[41] Among East Javanese in Sitiung 4,

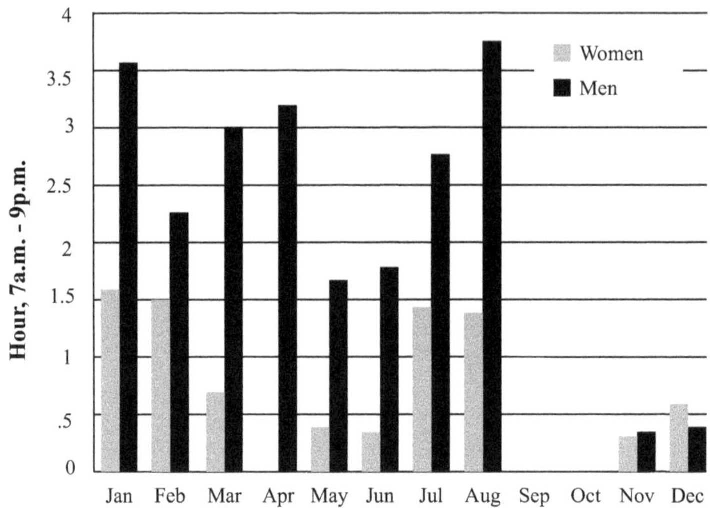

FIGURE 5.3 Seasonal variation in allocation of labour to upland fields by gender, Sitiung 1, 1983–1984.

Source: C. J. P. Colfer, *Toward Sustainable Agriculture in the Humid Tropics: Building on the TropSoils Experience in Indonesia*, TropSoils Bulletin No. 91–02, North Carolina State University, 1991.

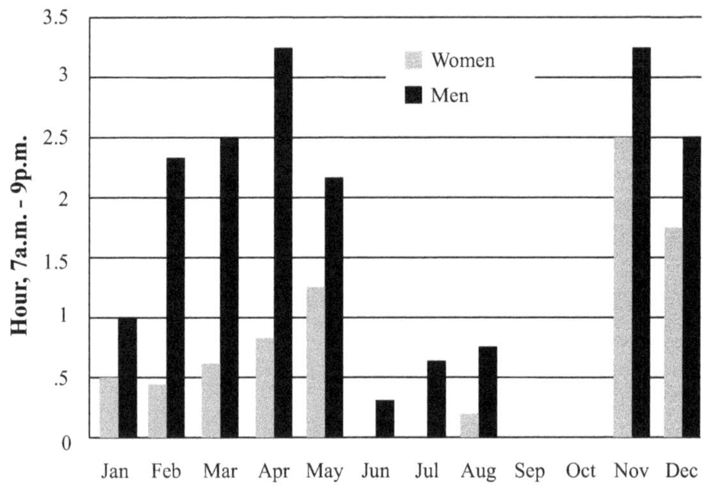

FIGURE 5.4 Division of labour by gender in paddy rice, Sitiung 1, 1983–1984.

Source: C. J. P. Colfer, *Toward Sustainable Agriculture in the Humid Tropics: Building on the TropSoils Experience in Indonesia*, TropSoils Bulletin No. 91–02, North Carolina State University, 1991.

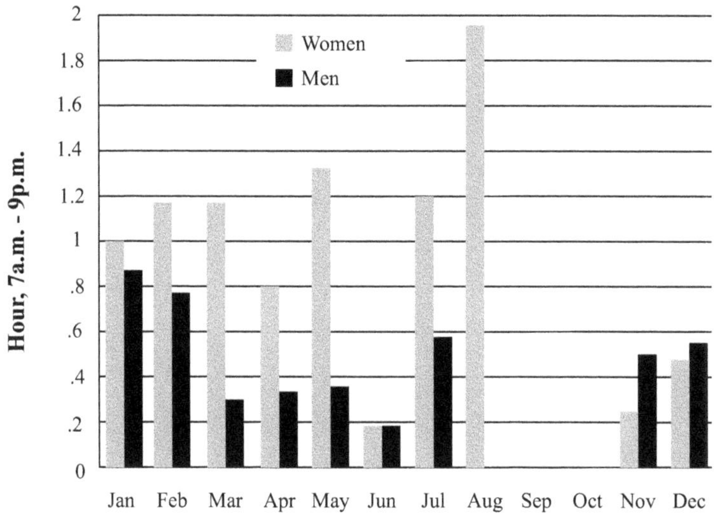

FIGURE 5.5 Division of labour by gender in home gardens, Sitiung 1, 1983–1984.

Source: C. J. P. Colfer, *Toward Sustainable Agriculture in the Humid Tropics: Building on the TropSoils Experience in Indonesia*, TropSoils Bulletin No. 91–02, North Carolina State University, 1991.

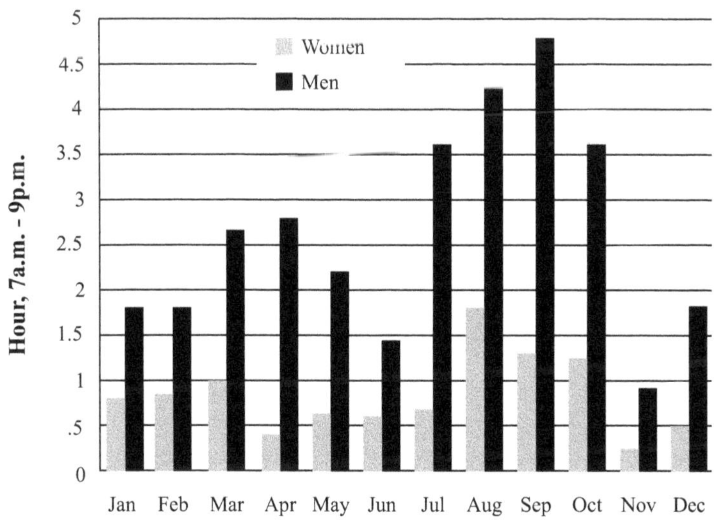

FIGURE 5.6 Division of labour by gender in upland fields, Sitiung 5, 1983–1984.

Source: C. J. P. Colfer, *Toward Sustainable Agriculture in the Humid Tropics: Building on the TropSoils Experience in Indonesia*, TropSoils Bulletin No. 91–02, North Carolina State University, 1991.

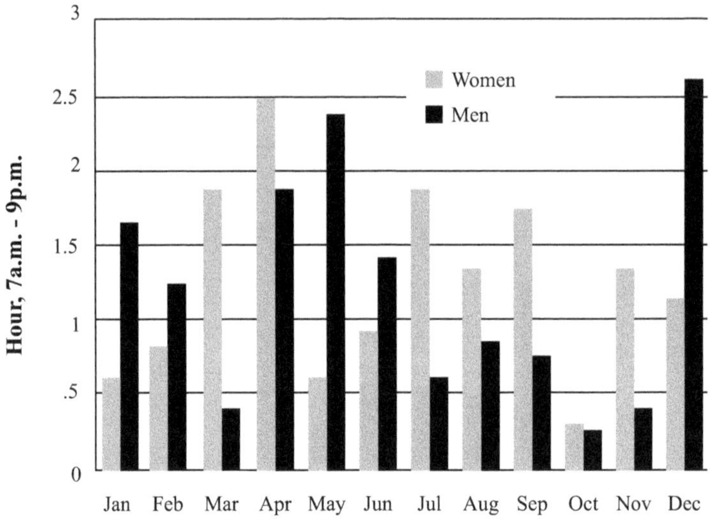

FIGURE 5.7 Division of labour by gender in home gardens, Sitiung 5, 1983–1984.

Source: C. J. P. Colfer, *Toward Sustainable Agriculture in the Humid Tropics: Building on the TropSoils Experience in Indonesia*, TropSoils Bulletin No. 91–02, North Carolina State University, 1991.

another transmigrant village, "No women helped with land preparation" (notes, 23 February 1984).

Expeditions, knowledge, violence and sports harp strings

Here we turn to four harp strings with more varying relevance in Sitiung. These harp strings pop up among some men, less consistently than those just discussed.

Expedition harp string

Minang men were famous nationally for their wide travels; to *merantau* means to go in search of adventure, knowledge and one's fortune (as among the Kenyah, Chapter 4). A common Indonesian joke at the time recognized this Minang stereotype: What was the first thing the astronauts found when they landed on the moon? Answer: A Padang restaurant (ubiquitous purveyor of Minang food). The Minang in this study, however, were somewhat disdained by those from the Minang heartland. Sitiung was considered part of the *rantau*, a pioneering area and thus seen as culturally somewhat impure.[42] The men I met had not travelled far, nor were there many others reported to have left home for distant places.[43] Naim and Hermann (1984) found only 3% of the men of Koto Padang away at the time of their 1984 survey. However, there was considerable local mobility, moving houses, moving villages, while remaining in the Sitiung area.

Although transmigrants who had come to Sitiung could be considered to *merantau*, these Javanese and Sundanese men did not refer to their move in that way. Rather, the motivation, if voluntary, was to 'change their fate' (*ganti nasib*), or improve themselves, and to obtain free land for farming. I never heard the idea of adventure—as often expressed by Americans, Kenyah and Minang from other regions—expressed.

Knowledge and magic harp strings

Local Minang men *did* comply with the stereotypical Minang love of education and knowledge. Although not well educated themselves, they had established and maintained a private religious boys' school in Pulai. They expressed intellectual interest in the studies I was conducting and a willingness, once they'd determined that I had no ulterior motive, to participate.[44]

Of the 40 families we interviewed in Sitiung 5 (project 'cooperator' and 'control' families), Minang men had the highest educational achievement, averaging 4.2 years of education, 1.7 more than Minang women. Although I do not have data on formal educational achievement in Pulai or Koto Padang, I suspect both are somewhat higher than this, as the poorest inhabitants were recruited to join the local transmigration programme. In contrast, East Javanese men averaged 2.7 years and Sundanese men, 3.8 years of education. In both these latter cases, the gender difference was smaller, only 0.2 years.

Minang men held many magical beliefs, including the belief that *sheytan* (Satan) and *Jihin* (djinn) caused many problems. People expressed a reluctance to go into the forest on Friday, reportedly because of an earlier experience in which 'some people' had done so and been eaten by tigers (notes, 3 January 1986).

> [My field assistant] said he'd gone halfway [to Pekanbaru, walking through the forest] himself seeking *barang kuno* (protective amulets) . . . said if you have one then a knife won't cut through your skin. But he didn't find one.
>
> *(notes, 23 October 1985)*

Kari, Pulai's old *dukun* (magical healer), when asked if he feared staying alone in his field hut in the rice field, responded that he had *ilmu* (magic/science), which protected him from tigers and other dangers (notes, 23 October 1985). He also blessed rice fields, with ginger and onions; a young Minang man felt that "no matter how long into the future, this custom would not disappear" (notes, 14 March 1986). Fear of tigers was a recurring concern. One woman sang while she tapped rubber to frighten away tigers and wild pigs (notes, 20 February 1986)—a downside of lush forests.

Although transmigrant men admired the educated and valued the availability of schooling in Sitiung 1, I did not sense the wide-ranging intellectual curiosity I saw among the Minang. Transmigrant men were, however, always interested to learn new agricultural techniques and try out new crops. Like the Minang, the transmigrants had elaborate magical beliefs about many aspects of human existence—though these were not as likely to pertain to the forest, agriculture or natural resources.

Violence harp string

Although violence was uncommon among the local Minang, on the most important religious festival, *Hari Raya* ('Id al Fitri'), little boys were seen (and heard!) exploding kerosene in 'guns' made from 6"-diameter bamboo from nearby forests. Although I only saw this once, this same bamboo was used to carry an alcoholic drink made from the "very healthful" (notes, 11 June 1986) *anau* (sugar palm, *Arenga pinnata*). A group of Minang men also expressed their shock and disapproval of widespread American gun ownership (notes, 9 May 1986).

I knew of several instances of wife-beating among the Javanese of Sitiung 1 and it seemed to be accepted by the community (if the woman had seriously transgressed) (notes, 3 Nov 1983; 26 November 1983; 17 December 1983). One of our drivers (half Javanese, half Madurese)—among the most macho men I met in Indonesia—liked the idea that his family was afraid of him. My next door neighbour and household helper was known to repeatedly stray sexually and ignore norms of women's 'appropriate comportment'. My driver was not fond of her, and said once apparently seriously, "If I'd been her husband, I'd have killed her".

On a subsequent occasion, this woman and another neighbour man were caught by her husband *in flagrante*; the husband badly beat the neighbour, but left his wife unscathed (notes, 4 January 1984). Indeed, he was reported to brag about his wife's attractiveness, as shown by other men's sexual interest in her.[45]

Another case of violence occurred among the Sundanese: In the early days of Sitiung 5 settlement, tensions ran high:

> Dody[46] himself was the one who beat up the transmigration official a long time ago, because of irregularities with *jata*.[47] After the beating . . . the official was honest and '*damai*' [peaceful], but someone else reported the incident to the police. This all came up because Omar . . . tried to kill Obing. Omar thought Obing was taking his *jata*, though Obing was getting three *jata*'s for other families. Someone grabbed the knife just in time, and Obing's wrist was only hurt, not his neck cut. The men were all at the mosque, as it was Friday . . .
>
> *(notes, 4 January 1984)*

The considerable inter-ethnic hostility between the Minang and in-migrating Javanese has been mentioned, but most played out in adverse stereotypes rather than violence per se. As Elmhirst (2018) observes in Lampung Province,

> pressures on livelihood and resources meant that what might otherwise be delight in diversity sometimes manifested itself through metaphors of

> moral disapproval about the practices of the other group [in this case the Javanese and the Lampungese]. . . . struggles over land and livelihood were frequently articulated through an idiom of cultural difference.
>
> *(p. 11)*

This was the case in Sitiung as well. Understanding these cultural differences—many of which relate to gender norms—can enhance our attempts at good forest management and related conflict management.

Sports and arts harp strings

In the US, sports are an important part of the formal educational system. In Pulai, a youth sports event was scheduled in connection with the boys' graduation. The fact that a group of Minang men expressed ignorance about the derivation of foreign words used in Indonesian soccer (keeper, out, offsides) reflected this sport's lack of centrality there. I heard once, during my last month in residence, that the people of Pulai sometimes had canoe races, though I heard no routine discussion of sports among them, nor did I see people engaged in sports there—a rarely plucked harp string.

Among the transmigrants in Sitiung 1, sports were more common. There were regular soccer games between the men on our research team and the village's young men. However, the funds to pay for uniforms and sports equipment (when not bought by the research team) was a constraint. Sports did not play the central role we found, for instance, in Bushler Bay.

Javanese men were more obviously involved in Javanese arts (gamelan, puppet shows, puppet making) than were Minang men in Minang arts.

Dominant masculinities summarized

Both local Minang and Javanese men were Muslim. The Minang were concerned with the day-to-day dictates of Islam (praying, fasting, avoiding pork, covering their heads, etc.); the rural Javanese, less so. The lack of Javanese attention to personal modesty (particularly women's) was a key bone of contention—linking, in Minang minds, such religious 'failings' to probable disapproved sexuality.

Both local Minang and in-migrating Javanese were intimately involved in their respective political hierarchies. Whereas the Minang men emphasized political action and entrepreneurship, at and above the local level, most Javanese men emphasized the need to comply with governmental directives, to submit to authority.

Men in both groups had provisioning responsibilities, but with different priorities: Minang men prioritized entrepreneurial income generation, with farming as a secondary choice; Javanese men emphasized farming, with earning supplementary cash of secondary import. Opportunities to enhance incomes came up serendipitously, which meant Minang men were often visible, sitting and waiting for such opportunities to emerge (as often happened). This pattern though, reinforced Javanese stereotypes of Minang laziness.

Hoeing and fertilizer use were seen by Javanese men as important for production, evidence of personal industriousness and symbolic of men's roles. Minang men were reluctant to hoe or fertilize, whereas Minang women hoed routinely in their care of paddy fields. The Javanese saw the Minang pattern as evidence of Minang men's laziness and an uncaring attitude towards their wives (a source of Javanese disdain). In fact, our project found that hoeing in that environment did not have a significant effect on yields.

Forests and conclusions

None of the main ethnic groups in Sitiung had the deep cultural narratives pertaining to forests that were so obvious among both the American loggers of Bushler Bay or the Kenyah of Kalimantan. Minang lifeways manifested a closer connection to the surrounding forest than did the in-migrating Javanese, who tended to fear the forest. Both the Javanese and the Sundanese were used to modes of making a living on Java that valued intensive cultivation of cash crops over subsistence agriculture—field crops in the Javanese case, with horticultural tendencies among the Sundanese, though in Sitiung 5, neither had sufficient excess to sell much.[48] This orientation largely removed their agricultural activities from direct forest connections.

The Minang in Sitiung depended on the availability of forest lands to grow upland rice in swiddens, and men and women of all ethnic groups were involved in rice production, both paddy and upland (see Figures 5.2 to 5.4 and 5.6)—a product the Minang considered shameful to sell. But the men particularly were more interested in the tree crops they planted soon after the rice, and which took over from the rice in subsequent years. The resulting fields of mostly rubber agroforestry (or 'jungle rubber') produced a landscape that looked like a forest to the uninitiated and provided much of the biodiversity and other environmental services extant in nearby secondary forests. It also required less day-to-day maintenance than conventional field crops.

Politics was a particularly significant part of Minang and Sundanese men's lives, and all three ethnic groups were far more intimately integrated into the lower reaches of the national political hierarchy than were either the Kenyah or Bushler Bay loggers. This integration held the potential for local people of these ethnic groups to influence at least provincial forest policy.

The different behaviours and beliefs related to masculinity can impinge importantly on forest management. The Javanese, the Sundanese and to a lesser degree, the Minang, are intimately interlinked with complex bureaucracies and the important tensions between submission and resistance that accompany such intimacy. Formal forest management occurs within such bureaucracies. The strong sense of duty to submit among many Javanese, for instance, can be an advantage for implementing forest management plans. On the other hand, such communities are likely to have developed sophisticated skills at subverting and resisting as well (à la Scott 1985).

Forest management can also be affected by bringing together groups with differing cultural preferences and practices—as is often done by governments seeking to 'rationalize' governance. In Indonesia, this has often included the inclusion of local peoples among Javanese transmigrants. Attempts, for instance, to banish shifting cultivation have included bringing in Javanese used to settled agriculture. This particular policy often resulted instead in *more* shifting cultivation in these areas of infertile soils and minimal access to agricultural inputs (e.g., Fulcher 1982). In no cases, to my knowledge, do such policies include mechanisms to cope with the resulting inter-ethnic conflict.

Governments have also ostensibly sought to provide improved infrastructure in remote areas, by resettling groups closer together (thus also 'liberating' forest-rich areas for lucrative concessions). Again, inter-group hostilities are likely to complicate forest management efforts, as forest managers cope with ongoing cultural and resource-related conflict (Peluso and Harwell 2001 document one of the more dramatic instances of inter-ethnic violence, in West Kalimantan, 1997).

Although I have not been back to Sitiung since 2008, at that time the landscape was already primarily oil palm, with some remnants of rubber cultivation in more orderly rows.[49] One concern, besides the loss of the forest (which has already happened), is the question of land tenure. Before the transmigrants arrived, the Minang communities had traditional land tenure that seemed reasonably secure. But in most parts of Indonesia's Outer Islands, land was also being allocated under President Soeharto to industries (timber concessions [HPH], *Hak Pengusahaan Hutan*; oil palm; industrial timber concessions [HTI] *Hutan Tanaman Industri*; mining; etc.), with little or no attention to pre-existing, local tenure rules. The Minang, with a broader base of educated people nationally, may have been able to protect their traditional lands better than average, though I do not know their degree of success.

This chapter has emphasized the differing concepts of masculinity, of men's (and women's) appropriate behaviour, so central in the antagonisms that beset inter-ethnic interactions in Sitiung. In Chapter 6, I turn away from formal research results, again drawing on my personal experience, as I examine masculinities among my colleagues in forest-related international conservation and development spheres.

Notes

1 Original material for this chapter comes from fieldnotes (in Cornell University's archives), memory and the following published works: Colfer (1991), Colfer, Newton, and Herman (1989), Colfer, Gill, and Agus (1988), Sigman et al. (1989a, b); and unpublished site reports, including Naim and Hermann (1984). The harp produced here is a likely example of the harp strings an imaginary Minang man might pluck from time to time.

2 I occasionally use pseudonyms in this chapter.
3 Sitiung included at that time five numbered areas, each including smaller hamlets identified by letter, interspersed with local, long-resident Minangkabau ('Minang') communities. We called the hamlets where we worked Sitiung 1 and 5, as I do here.
4 Between 1976 and 1986, some 6000 transmigrant families moved to the area (Colfer 1991), p. 10); more came later.
5 The dominant ethnic group in West Java.
6 Transmigrants typically applied to join such a programme and were sent to areas where they had no pre-existing community relations (see Elmhirst 2011, p. 177, for a gender-sensitive discussion of this programme).
7 Although all three Sitiung ethnic groups represent 'lowlanders' in Scott's (2009) sense (similar to the Makassar Gibson studied), this differentiation does not accurately represent the actual homelands or current geographies of the groups I studied; the Minang, though dominant in West Sumatra, lie somehow between the upland and lowland 'ideal types'. The Minang in (lowland) Sitiung also bring to mind Li's (1999a) warning about the simplifications and reifications of Indonesian ethnic groups (along with related political implications), some of which have very fuzzy boundaries. The Sitiung Minang, for instance, seem to have been somewhat influenced by the more patrifocal Jambi groups to the southeast.
8 See Jones (2019), for a discussion of the frequency of these issues in Indonesia.
9 Elmhirst (2018) discusses a similar context in southern Sumatra:

> In transmigration, where Javanese transmigrants and Lampung people were brought together, contestations of meaning and identity were most intense in the space between the two communities and struggles over land and livelihood were frequently articulated through an idiom of cultural difference. These struggles rarely took on an overt physical or violent form. Rather, by far the most common way in which the two groups interacted and challenged each other's legitimacy was through the hidden and sometimes not so hidden "transcripts" (Scott 1990)—gossip and rumour—through which one group represented and resisted the actions and representations of the other.
>
> *(p. 11)*

See also Algiovan, Umami, and Amanah (2016) who attribute the sources of ethnic disagreements and conflict in East Lampung to "economic issues . . . theft, spoliation, and violence", as well as low educational levels and poverty (p. 560).
10 Matrilineality does *not* imply a matriarchate (i.e. that women hold the political power). Formal political and clan power were held by women's brothers. Still, the fact that certain kinds of land ownership followed the female line granted women powers that were less available to women in a patrilineal society. See Blackwood (1995) for a good discussion of a Minang system in their heartland.
11 The historical process of combining the two was not smooth. See Dinata's (2013) brief account or McKay (2013), who examines the history of Islam and its integration with custom in West Sumatra.
12 In contrast to this overt religiosity, few Bushler Bay people attended church, and far fewer men than women. I do not remember a Bushler Bay man ever mentioning religion positively in the 1970s. Many considered the nearby Pentecostal 'Faith Farm' amusing. In the international professional community (Chapter 6), there was a similar reluctance to profess any religious beliefs or practices. Highly educated Americans also seem reluctant to profess adherence to formal religious beliefs publicly.
13 In this study, 21 locally relevant concepts about agriculture were paired and their perceived distance from each other measured by 100 respondents from each of the three ethnic groups. We used a 'measuring stick' of each respondent's own cognitive distance between *black* and *white*, set at 100 units apart (see Colfer et al. 1989, Colfer 1991).
14 Minang respondents considered these agricultural and soil management concepts overall to be an average of 44 units apart, in contrast to the more agriculturally oriented Javanese (mean of 30) and Sundanese (mean of 22; Colfer et al. 1989).

15 *Ilmu* also means science.
16 In contrast, magic used in such a case among the Kenyah would have resulted in her falling in love with the perpetrator (as I was told happened to me).
17 No gender differentiation in pronouns in any of these languages.
18 In the Minang heartland, the woman's family proposes marriage (e.g., Fahmi and Aswirna 2014).
19 There were exceptions: When a Javanese widow in Sitiung 1 was in desperate financial straits, she arranged a marriage for her reluctant 13-year-old daughter, despite general disapproval.
20 Three, year-long, observational time allocation studies were conducted in Sitiung 1, 5 and Pulai, based on 5634 observations, showing considerable ethnic and gender variation. The data from Sitiung 1 involved cyclical visits (date and time of day throughout the year) to Central Javanese homes; from Pulai, to Minang homes; and from Sitiung 5 to East Javanese, Sundanese and Minang homes—all covering the entire settlements.
21 Although here, I emphasize the conflicts among the Indonesian residents, there were also conflicts between the residents and our research team members (Box 5.2). Politics and religion are closely linked for many in Indonesia.
22 Matrilineal systems have often confused broader-scale actors. Davison (1997) discusses the problems created by colonial powers in southeastern Africa whose representatives considered matrilineality to characterize an earlier stage in human evolution. Spall (2016) describes the Christian church's preaching against matrilineal inheritance in Angola, concluding,

> Indeed, fathers' responsibility for their children seemed to be growing, with the influence of maternal uncles declining in the face of state legislation limiting inheritance to spouses and children (rather than siblings, nephews and nieces) and churches preaching against matrilineal inheritance.
>
> (*p. 157*)

See also Kato (1982) for good discussion of matriliny's capacity to endure in West Sumatra.
23 Peletz (1995) notes that "Malay societies typically gloss over the strong emphases on matrifocality . . . and matrilaterality characteristic of all Malay systems of social relations" (p. 81), as in Indonesia.
24 The fourth clan head was away making money.
25 The highest official I ever saw in Long Segar, which admittedly was more remote, was the *Camat*, once. Extra-local officials in East Kalimantan were almost always Javanese men, appointed by higher-level government officials.
26 Cf. Long Segar, where both young women and men served on formal occasions.
27 Kato's (1982) work suggests that these folks may have initially migrated to this area in the 17th century (though possibly as early as the 15th, p. 112) in search of gold—a resource women still panned for in the Batang Hari River.
28 See Kahin (1999) for a careful historical analysis of West Sumatran administrative structures and Vel and Bednerb (2015) for a nice update on how the Minangkabau have creatively adapted and made use of the changes wrought after Indonesia's 2001 decentralization, returning desired aspects of the earlier *nagari* system of local governance. Vel and Bednerb mention also the "triangle of state-*adat* [custom]-religion" (p. 496)—evident in the mid-1980s as well: the elements I chose for the Minang harp itself.
29 This refers to communities that were moved en masse, in this case the area now inundated by the Gajah Mungkur dam.
30 People answered this question variably: Some people were extremely poor on Java and felt wealthy with the minimal amenities available in Sitiung; others, particularly those who moved to make way for the Wonogiri dam or to escape Galunggung's eruption, had been better off on Java.
31 Such submissiveness was also common among Javanese women, who accepted family planning required by official policy with the same spirit.
32 Consider the strong norm on sharing (and generalized reciprocity) among the Kenyah and the sort of 'Image of Limited Good' (Foster 1965) reflected here in this context

once termed 'peasant'. Thanks to Michael Cummings for reminding me of this, in different terms.
33 This *suku* is Minang for 'clan', unlike the Indonesian cognate meaning 'ethnic group'.
34 This may have reflected cultural influence from the nearby patrilineal Jambi ethnic group, reminiscent of the ethnic continua emphasized by Kahn (1999) and the ethnic ambiguity discussed in Li (1999b).
35 When clan decisions were being made however, the men were not so marginalized.
36 Davison (1997) reminds us that marginalization equally characterizes in-marrying women in patrilineal groups.
37 E.g., coffee, rambutan, bananas, stink bean, coconut, lemon, papaya—planted among rubber trees and in home gardens.
38 On 8 January 1986, I weeded with Minang women for two hours while my assistant sat watching us, never lifting a finger, answering my questions and translating as needed. I did see one old man weeding and one handicapped man harvesting, on other days.
39 Koto Padang was a *jorong* (administrative district) of the *Kenagarian* Sialang Gaung, in Sawahlunto/Sijunjung.
40 Similar inaccurate self-deprecatory remarks were also common among the Kenyah, in recognition of the uncomplimentary views other ethnic groups held of them.
41 During the first year of habitation, 1983–1984, the people were busy with all the myriad tasks involved in making their homes and community livable. They also were able to survive on the government rice subsidies during this first year.
42 A different perception from American notions of the frontier—a rather central meme in white American ideology.
43 Kato (1982) also noted the lower rates of expedition-making in the *rantau*, with an estimate of 8% in Sawahlunto-Sijunjung, the Kabupaten in which Sitiung was located (p. 143).
44 False rumours about one of our team members' purported religious activities (see Box 5.2) heightened local tension about my own motivations. There were many local beliefs about the tricks Christians purportedly played to gain converts. My help to a woman in danger of dying in childbirth was a major help in overcoming community reservations (Colfer 1992). Initially the level of suspicion was high, as it had remained in Bushler Bay, where there was less intellectual interest in our work.
45 Lending some credence to this interpretation relating to the Sundanese: "'Although men may appear to be competing for the women, more significant is their jousting with other men for prestige' (Cooper 2000, p. 627)" [quoted in Spiller 2010, p. 40].
46 Dody and Omar are Sundanese; Obing's ethnicity is not recorded.
47 *Jata* is the benefits provided to transmigrants by the government, including monthly food and agricultural equipment, during the first year (1983). The former was very important in Sitiung 5, where crops had already failed when we arrived.
48 Wessing (1999) seeks evidence of a strong love of forests in West Java without much success (except historically and among some remnant populations like the Baduy and the Kasepuhan). The collection by McKay (2013) makes a more convincing argument for the role of nature and forests among the Minang.
49 www.forestpeoples.org/en/palm-oil-rspo/press-release/2019/press-release-report-shows-widespread-human-rights-violations: This website, which includes discussion of West Sumatra's problems with the oil palm giant, Wilmar, suggests problems similar to those found in East Kalimantan (Chapter 7).

References

Algiovan, Nyanuar, Ida Umami, and Binti Amanah. 2016. "The Role of Citizens in Increasing the Quality of Multicultural Education (A Micro Ethnographic Studies in East Lampung Regency)." *Batusangkar International Conference I*, 15–16 October.

Barker, Gary, Marcos Nascimento, Christine Ricardo, Marianna Olinger, and Márcio Segundo. 2011. "Masculinities, Social Exclusion and Prospects for Change:

Reflections from Promundo's Work in Rio de Janeiro, Brazil." In *Men and Development: Politicizing Masculinities*, edited by Andrea Cornwall, Jerker Edström, and Alan Greig, 170–184. London: Zed Books.

Blackwood, Evelyn. 1995. "Senior Women, Model Mothers, and Dutiful Wives: Managing Gender Contradictions in a Minangkabau Village." In *Bewitching Women, Pious Men: Gender and Body Politics in Southeast Asia*, edited by Aihwa Ong and Michael G. Peletz, 124–158. Berkeley: University of California Press.

Brenner, Suzanne A. 1995. "Why Women Rule the Roost: Rethinking Javanese Ideologies of Gender and Self-Control." In *Bewitching Women, Pious Men: Gender and Body Politics in Southeast Asia*, edited by Aihwa Ong and Michael G. Peletz, 19–50. Berkeley: University of California Press.

Cassel, D. Keith, A. Karim Makarim, and Michael K. Wade. 1987. "Reclamation of Bulldozed Lands." In *TropSoils Technical Report*, edited by Neil Caudle and Charles B. McCants. Raleigh, NC: North Carolina State University.

Caudle, Neil, and Charles B. McCants. 1987. *TropSoils Technical Report, 1985–86*. Edited by Neil Caudle and Charles B. McCants. Raleigh, NC: North Carolina State University.

Colfer, Carol J. Pierce. 1981. "Women, Men and Time in the Forests of East Kalimantan." *Borneo Research Bulletin* 13 (2):75–84.

Colfer, Carol J. Pierce. 1991. *Toward Sustainable Agriculture in the Humid Tropics: Building on the Tropsoils Experience in Indonesia. Vol. No. 91/02, Tropsoils Technical Bulletin*. Raleigh, NC: North Carolina State University.

Colfer, Carol J. Pierce. 1992. "Of Teamwork, Faith, and Trust in West Sumatra." In *The Naked Anthropologist: Tales from Around the World*, edited by Philip R. DeVita, 90–99. Belmont, CA: Wadsworth Publishing Company.

Colfer, Carol J. Pierce, Ramadhani Achdiawan, James M. Roshetko, Elok Mulyoutami, E. Linda Yuliani, Agus Mulyana, Moira Moeliono, Hasantoha Adnan, and Erni. 2015. "The Balance of Power in Household Decision-Making: Encouraging News on Gender in Southern Sulawesi." *World Development* 76:147–164.

Colfer, Carol J. Pierce, Dan Gill, and Fahmuddin Agus. 1988. "An Indigenous Agroforestry Model from West Sumatra: A Source of Insight for Scientists." *Agricultural Systems* 26:191–209.

Colfer, Carol J. Pierce, Dan Gill, Barbara J. Newton, and Herwasono Soedjito. 1989. "Indigenous Knowledge Systems Related to Soil Management." In *TropSoils Technical Report*, edited by Neil Caudle. Raleigh, NC: North Carolina State University.

Colfer, Carol J. Pierce, B. Newton, and Herman. 1989. "Ethnicity: An Important Consideration in Indonesian Agriculture." *Agriculture and Human Values* VI (3):52–67.

Colfer, Carol J. Pierce, Russell Yost, Veronica Kasmini, and Atin Kurdiana. 1984. "Time Allocation Studies: A Methodology for Food Production Systems." *World Food Day Symposium on Food Production Systems*, Ithaca, New York, 16 October.

Davison, Jean. 1997. *Gender, Lineage and Ethnicity in Southern Africa*. Boulder, CO: Westview Press.

Dewey, Alice. 1962. *Peasant Marketing in Java*. New York: Free Press of Glencoe.

Dinata, Yoan. 2013. "The History of the Minangkabau." In *Integrating Religion Within Conservation: Islamic Beliefs and Sumatran Forest Management*, edited by Jeanne McCay, 21–23. Canterbury, Kent: Durrell Institute of Conservation and Ecology (DICE).

Elmhirst, Rebecca. 2011. "Migrant Pathways to Resource Access in Lampung's Political Forest: Gender, Citizenship and Creative Conjugality." *Geoforum* 42:173–183.

Elmhirst, Rebecca. 2018. "Departing from Java to Lampung: Locating Javanese Diasporic Practices in Indonesia's Transmigration Program." In *Departing From Java:*

Javanese Labour, Migration and Diaspora, edited by Rosemarijn Hoefte and Peter Meel. Copenhagen: NIAS Press.

Fahmi, Reza, and Prima Aswirna. 2014. "The Paradox of Islam and Culture (Tradition and Belief about Gender Perspective in West Sumatra)." *Al Turas: Mimbar Sejarah, Sastra, Budaya dan Agama* XX (2):67–75.

Firmansyah, Nurul. 2013. "Part II—Islamic Teachings and Traditions in Community-Based Forest Management: A Case Study in Nagari Guguk Malalo." In *Integrating Religion Within Conservation: Islamic Beliefs and Sumatran Forest Management*, edited by Jeanne E. McKay, 41–51. Canterbury, Kent: Durrell Institute of Conservation and Ecology (DICE).

Foster, George. 1965. "Peasant Society and the Image of Limited Good." *American Anthropologist* 67 (2):293–315.

Fulcher, Mary. 1982. "Dayak and Transmigrant Communities in East Kalimantan." *Borneo Research Bulletin* 14 (1):14–24.

Geertz, Clifford. 1963. *Agricultural Involution: The Process of Ecological Change in Indonesia*. Berkeley, CA: University of California.

Gibson, Thomas. 2005. "From Humility to Lordship in Island Southeast Asia." In *Property and Equality (Volume II: Encapsulation, Commercialisation, Discrimination)*, edited by Thomas Widlok and Wolde Gossa Tadesse, 231–251. New York: Berghahn Books.

Gibson, Thomas. 2011. "Egalitarian Islands in a Predatory Sea." In *Anarchic Solidarity: Autonomy, Equality and Fellowship in Southeast Asia*, edited by Thomas Gibson and Kenneth Sillander, 270–293. New Haven: Yale University Southeast Asia Studies Monograph.

Gill, Dan W., Sri Adiningsih, and Antonius Kasno. 1987. "Potassium Management of Upland Crops." In *TropSoils Technical Report*, edited by Neil Caudle and Charles B. McCants. Raleigh, NC: North Carolina State University.

Gouyon, A., H. De Foresta, and P. Levang. 1993. "Does 'Jungle Rubber' Deserve Its Name? An Analysis of Rubber Agroforest Systems in Southeast Sumatra." *Agroforestry Systems* 22:181–206.

Hall, Maurice. 2011. "Negotiating Jamaican Masculinities." In *Global Masculinities and Manhood*, edited by Ronald L. Jackson II and Murali Balaji, 31–51. Urbana, IL: University of Illinois Press.

Jakimow, Tanya. 2017. "Beyond 'State Ibuism': Empowerment Effects in State-led Development in Indonesia." *Development and Change* 49 (5):1143–1165. https://doi.org/10.1111/dech.12374.

Jones, Sidney. 2019. "Manipulating Minorities and Majorities: Reflections on 'Contentious Belonging'." In *Contentious Belonging: The Place of Minorities in Indonesia*, edited by Greg Fealy and Ronit Ricci, 255–265. Singapore: ISEAS Yusof Ishak Institute.

Kahin, Audrey. 1999. *Rebellion to Integration: West Sumatra and the Indonesian Policy 1926–1998*. Amsterdam, The Netherlands: Amsterdam University Press.

Kahn, Joel S. 1999. "Culturalising the Indonesian Uplands." In *Transforming the Indonesian Uplands: Marginality, Power and Production*, edited by Tania Li, 79–104. Amsterdam, Netherelands: Harwood Academic Publishers.

Kato, Tsuyoshi. 1982. *Matriliny and Migration: Evolving Minangkabau Traditions in Indonesia*. Ithaca, NY: Cornell University Press.

Krier, Jennifer. 1995. "Narrating Herself: Power and Gender in a Minangkabau Woman's Tale of Conflict." In *Bewitching Women, Pious Men: Gender and Body Politics in Southeast Asia*, edited by Aihwa Ong and Michael G. Peletz, 51–75. Berkeley: University of California Press.

Laderman, Carol. 1996. "Putting Malay Women in Their Place." In *Women of Southeast Asia*, edited by Penny Van Esterik, 62–77. DeKalb, IL: Northern Illinois University.

Lentz, Linda. 2017. *The Compass of Life: Sundanese Lifecycle Rituals and the Status of Muslim Women in Indonesia*. Durham, NC: Durham Academic Press.

Li, Tania. 1999a. "Marginality, Power and Production: Analysing Upland Transformations." In *Transforming the Indonesian Uplands: Marginality, Power and Production*, edited by Tania Li, 1–46. London: Taylor and Francis e-Library.

Li, Tania, ed. 1999b. *Transforming the Indonesian Uplands: Marginality, Power and Production*. Amsterdam, Netherlands: Harwood Academic Publishers.

McKay, Jeanne E., ed. 2013. In *Integrating Religion Within Conservation: Islamic Beliefs and Sumatran Forest Management*. Canterbury, Kent: Durrell Institute of Conservation and Ecology (DICE).

Michon, Genevieve, F. Mary, and J. Bompard. 1986. "Multistoried Agroforestry Garden System in West Sumatra, Indonesia." *Agroforestry Systems* 4:315–338.

Naim, Mochtar, and Hermann. 1984. *Tree Crop Farming System Pattern in Koto Padang, Sawahlunto/Sijunjung District, West Sumatra*. Sitiung, West Sumatra, Indonesia: TropSoils (University of Hawaii and Indonesian Center for Soil Science).

Peletz, Michael G. 1995. "Neither Reasonable nor Responsible: Contrasting Representations of Masculinity in a Malay Society." In *Bewitching Women, Pious Men: Gender and Body Politics in Southeast Asia*, edited by Aihwa Ong and Michael G. Peletz, 76–123. Berkeley, CA: University of California Press.

Peluso, Nancy Lee, and Emily Harwell. 2001. "Territory, Custom and the Cultural Politics of Inter-Ethnic War in West Kalimantan, Indonesia." In *Violent Environments*, edited by Nancy Lee Peluso and Michael Watts, 83–116. Ithaca, NY: Cornell University Press.

Scott, James C. 1985. *Weapons of the Weak: Everyday Forms of Peasant Resistance*. New Haven, CT: Yale University Press.

Scott, James C. 1990. *Domination and the Arts of Resistance: Hidden Transcripts*. New Haven: Yale University Press.

Scott, James C. 2009. *The Art of Not Being Governed: An Anarchist History of Upland Southeast Asia*. New Haven, CT: Yale University Press.

Sigman, Vickie A., Carol J. Pierce Colfer, Kathleen Wilson, and Russell Yost. 1989a. "Indonesia: Farm-Based Research in the Tropsoils Project, Sitiung." In *Working Together: Gender Analysis in Agriculture*, edited by Hilary Sims Feldstein and Susan V. Poats, 135–164. Harford, CT: Kumarian Press.

Sigman, Vickie A., Carol J. Pierce Colfer, Kathleen Wilson, and Russell Yost. 1989b. "Indonesia: Part 2, Research in the Second and Third Years." In *Working Together: Gender Analysis and Agriculture*, edited by Hilary Sims Feldstein and Susan V. Poats, 160–166. Hartford, CT: Kumarian Press.

Siskind, Janet. 1973. *To Hunt in the Morning*. London: Oxford University Press.

Spall, John. 2016. "Fatherhood and Intergenerational Struggles in the Construction of Masculinities in Huambo, Angola." In *Masculinities Under Neoliberalism*, edited by Andrea Cornwall, Frank G. Karioris, and Nancy Lindisfarne, 151–164. London: Zed Books.

Spiller, Henry. 2010. *Erotic Triangles: Sundanese Dance and Masculinity in West Java*. Chicago, IL: University of Chicago Press.

Suyanto, S., Thomas Tomich, and Kijiro Otsuka. 1997. "The Role of Land Tenure in the Development of Cinnamon Agroforestry in Kerinci, Sumatra." In *Voices from the Forest: Integrating Indigenous Knowledge into Sustainable Upland Farming*, edited by Malcolm Cairns, 743–753. Washington, DC: Resources for the Future.

Tanner, Nancy. 1974. "Matrifocality in Indonesia and Africa and Among Black American." In *Women, Culture and Society*, edited by Michelle Zimbalist Rosaldo and Louise Lamphere, 129–156. Stanford, CA: Stanford University Press.

Vel, J. A. C., and A. W. Bednerb. 2015. "Decentralisation and Village Governance in Indonesia: The Return to the Nagari and the 2014 Village Law." *The Journal of Legal Pluralism and Unofficial Law* 47 (3):493–507. http://doi.org/10.1080/07329113.2015.1109379.

Wade, Michael K., Heryadi, and Dan W. Gill. 1987. "Effect of Green Manure Applications on Soil Fertility and Crops." In *TropSoils Technical Report*, edited by Neil Caudle and Charles B. McCants. Raleigh, NC: North Carolina State University.

Wade, Michael K., Eugene J. Kamprath, Heryadi, Al-Jabri, Edi Joniarta, and Fahmuddin Agus. 1987. "Liming in Transmigration Areas." In *TropSoils Technical Report*, edited by Neil Caudle and Charles B. McCants. Raleigh, NC: North Carolina State University.

Wade, Michael K., Djoko Santoso, Al-Jabri, and I. Putu Gedjer. 1987. "Phosphorous Management in Transmigration Areas." In *TropSoils Technical Report*, edited by Neil Caudle and Charles B. McCants. Raleigh, NC: North Carolina State University.

Wessing, Robert. 1999. "The Sacred Grove: Founders and the Owners of the Forest in West Java, Indonesia." In *L'Homme et la Forêt Tropicale*, edited by Serge Bahuchet, Daniel Bley, Hélène Pagezy, and Nicole Vernazza-Licht. Marseille, France: Société d'écologie humaine.

Whitten, Anthony J., Sengli J. Damanik, Jazanul Anwar, and Nazaruddin Hisyam. 1984. *The Ecology of Sumatra*. Yogyakarta, Indonesia: Gadjah Mada University Press.

6

MASCULINITIES AMONG 'THE DEVELOPMENT SET'[1]

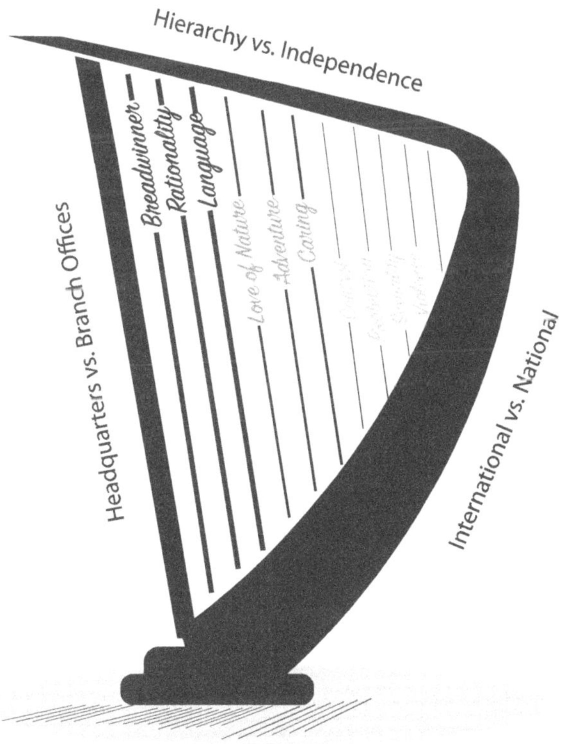

An International Scientist
Harp Example

Introduction to masculinities at work

In this chapter, I turn to the men (and women) with whom I have worked during my professional life. The analyses differ from those in Chapters 3 to 5 and 7, in relying on my personal experience without formal study (cf. Katila and Meriläinen 1999). For that reason, readers must take account of my own 'standpoint' (à la Harding 2004) more fully.[2] I share many of the ideals and values held by my colleagues (at least in the work setting), the most notable related to the muting of gender difference.

This analysis strives to understand the ideas that elite men in international forest-related conservation and development [appear to] have about their own masculinities. 'Studying up', as so many have urged us to do, has proven difficult and rare.[3] It seems probable that structural issues that we know interfere with equality in our work settings are related to masculinities in unclear ways. Perhaps by analyzing my experience in settings widely deemed masculine, I can shed some light on these dynamics and their broader implications.

One factor that has made this process difficult is identified by Hearn (2011) and others; he notes that

> much of what men do is not seen as 'about gender'. . . . Most of their practices, in public and in private – in work, negotiations, persuasion, networking, lobbying, pressurizing and so on – are neither seen nor experienced as gendered. They are done, perceived and felt as (if they were) 'normal'.
>
> *(p. 159)*

And women, like myself, who have succeeded reasonably well in such masculine contexts, have taken on some of these same behaviours, attitudes and skills that were traditionally associated with men, complicating the task of differentiation still further. Yet . . . I plunge into the abyss.

I focus here primarily on masculinities in the work setting, with a short addendum on related home settings. Harp strings used and valued at work may differ from those used and valued at home, though as Katila and Meräinen (2002) point out, this boundary can be porous; it can also be particularly drastic in these international settings. International work settings include inequities with widespread effects, including cognitive dissonance for those whose actions reinforce the inequities (typically unintentionally; examples below).

In the coming analysis, I have tried to disguise individuals insofar as I am able—many injustices are created and reinforced by structural features beyond individual control. I avoid personal names or invent pseudonyms; I use pseudonyms for the main organizations and projects; I use plurals when reality may have been singular; and I disguise dates (e.g., 199X), for further anonymity—all in an attempt to ensure anonymity. My feelings for my colleagues are overwhelmingly positive, as are my perception of their benign intentions and my own unwitting collaboration in such injustices.[4] Although I am committed to presenting this analysis in as intersectional a manner as possible, it has been necessary in maximizing anonymity, to leave out some details of age and ethnicity that could be significant.

I also recognize that two of the three institutions I discuss here have been far better than average, in terms of equity. All three have recognized the importance of attending to both biophysical and social elements of the worlds we have struggled to improve (Li 2007). Although I genuinely appreciate all the people whose behaviour I appear to criticize in the pages to follow, I have tried to analyze my experiences truthfully and as gently as I am able—in search of behaviour and values linked to masculinities at work.

The observations below take place in the office, in travel, in the field and in international meetings. The participants come from cultural systems from all over the world. So the masculinities that emerge in this breadwinning site, the harp strings that are plucked there, represent a particular segment of life rather than the totality of a given culture. One would expect that chords that include the breadwinning harp string might predominate, and that varying cultural systems (or harps) embodied in the different ethnicities present may clash and create discordant sounds. Still, I've found that some commonalities are identifiable.[5]

Connell's early (1995) book on masculinities discussed four groups of Australian men, two of which are relevant for this work: one, the most relevant, she labelled 'Men of Reason'; the other examined men involved in the environmental movement. The characteristics identified for these two groups have been the emphases of most of the men I've worked with, though age and historical period have impacts as well, as I try to show.

The harp strings that emerge in my own experience include rationality, control, love of nature, quantification, articulateness, adventure, caring and protection and sexuality and violence—in no particular order. I describe three different harp frames, reflective of different social contexts and times, though whether this analogy can legitimately be expanded to such work settings is for the reader to judge. The harp example shown by the title of this chapter is from the Forest Research Institute (FRI).

I present these observations chronologically, so we can keep in mind the historical trends in society, the dynamics of contexts, while maintaining awareness of the possible implications of changes in the observer herself (my own aging and experience). The amount of detail increases below as the time frame shortens.

I begin with memories of my early professional life in Bali and East Kalimantan, as a solitary scholar in my early 30s. I then describe the professional views and behaviour of my colleagues on the project wherein I conducted the research reported in Chapter 5 (Sitiung, West Sumatra). I differentiate the masculinities that characterized the Americans on the team and those that related to interactions between Americans and Indonesian personnel (almost all men).

The third section briefly describes the masculinities performed in the management of a conservation project in West Kalimantan. It presents conflicting views of masculinity between conservation-oriented Dutch men and business-oriented Scots, as demonstrated in the management of a project I worked on in remote Danau Sentarum Wildlife Reserve (now National Park), West Kalimantan in the early 1990s.

The fourth section is the most elaborate and based on two decades of shared work experience. It discusses masculine harp strings at FRI (headquartered on Java), in nine sub-sections, concluding with a discussion of the harp itself (insofar as that analogy remains suitable). I conclude with a summary of the findings and some thoughts on their implications.

The early years

What comes to mind when I think back on my earliest professional international work (initially in Bali and then East Kalimantan) is the centrality of the sexuality harp string (analyzed more fully by Katila and Meräinen 2002 in a more recent Finnish academic context). I was a young woman in my early 30s and alone; men's sexuality (and my own) emerged quickly as issues I would have to address. The following provide a few representative experiences related to sexuality and masculinities:

- One middle-aged American colleague in East Kalimantan, once we'd finalized our research proposal in 197X, made his sexual interest clear, later suggesting we marry—explicitly to make a sexual relationship acceptable in the field. He was also a good mentor, articulate, intelligent, competitive and controlling. He valued rationality, perhaps above all else.
- In Bali, an important administrator in the medical school flirted with me consistently (personal journal, 14 August 197X), but also plucked the caring harp string, by helping me with medical problems and with my research and expressing care and respect for the village people he served. He was also articulate, intelligent and skilled at networking and politics (see Chapter 4).
- The husband—a Balinese doctor and grad student in the US—of my hostess in Denpasar kindly arranged for me to stay with her. He also insisted that his female relatives sleep outside her bedroom to protect her sexual virtue while he was away (notes, 18 July 197X; Chapter 4).
- My 1980s sexual relationship with my same-aged Kenyah field assistant in Long Segar has already been described (Chapter 1)—initiated by him, but also serving to protect me from some advances by others. My protection from others' sexuality was not his concern, however; he considered it my responsibility to discourage such advances (Chapter 4).
- I was harangued in the early 1980s, as I waited in the Samarinda (East Kalimantan) phone office with lewd comments by a middle-aged Australian forester, who clearly saw me as 'prey'.

- In the early 1980s, I was invited for a weekend in Singapore with Joe, an American logger from the nearby timber concession.[6] He assured me that it need not include sex, though my skepticism led me to refuse (see the next paragraph).
- In a 198X personal journal, I encountered reference to an attempt to seduce me by a Hungarian in Bangkok (22 June), another by a German in Pontianak (11 July) and a Swede, who offered to marry me, stay home and do childcare in Padang. Heterosexual opportunities were regularly on offer in those days from professional men I met, while away from home.

Joe was from the Pacific Northwest and he and his colleagues exhibited many of the work behaviours described for loggers in Bushler Bay (Chapter 3). Among themselves, the loggers behaved in very egalitarian ways, valuing independence, hard physical labour, physical strength and skill with heavy equipment (cf. Hendriks 2014, 2017 for European men in a logging concession in the Democratic Republic of Congo). These Americans in Indonesia operated within what was essentially a caste system. The American loggers (and wives) had American suburban-style housing atop a hill with American food flown in regularly. They interacted with Indonesians at work (the log dump, in the woods, etc.). Mid-way up the hill were clustered Filipinos, brought in to serve as foremen and in middle management. At the bottom of the hill, near the pre-existing Kutai village, were the Indonesian labourers—most brought in from other islands—housed in one-room, unfurnished, wooden apartments in 'longhouses'. There was also housing for prostitutes, reportedly mainly from Java. Elite Javanese counterparts, who rarely visited, predominantly dealt with the Americans, resenting any intrusion by the Kenyah, for instance, into the American space. My Kenyah colleague was totally ignored by the visiting official in a conversation I initiated, intended to be among the three of us; the hierarchically oriented official considered my colleague beneath his notice. I mention this setting in more detail than the others above, because it previews, to some degree, the conflicts that exist in subsequent international contexts: a clash of both social structures and masculinities that are discordant and uncomfortable for all.

In all these examples[7] from the 1970s and '80s—surely familiar to many older readers—sexuality and men's sexual initiative (though often enjoyed by both genders) were linked with ideas about how men should and could behave—some representing the popular 'bad boy' image (Chapter 5).

During this time, my self-confidence blossomed. I was treated with great respect by Indonesian academic men at Mulawarman University, a response I had not encountered either as a graduate student or young professional in the US—very pleasant. But I was also aware (and embarrassed) that my skin colour and PhD, rather than my articulateness, rationality, intelligence or any other quality I might have valued in myself, were the bases on which this overt respect was granted.

American masculinities in Honolulu, Hawaii and Sitiung, West Sumatra, Indonesia (1981–1986)

Concerns about justice and equity have permeated my work contexts. My job in West Sumatra emerged out of an Island University committee (all men, but me), which I led with a colleague looking at Farming Systems Research and Development (FSR&D) as a potentially fruitful way to approach international agriculture. Although still single, I was not subjected to sexual advances by these men, who ranged in age from late 30s to late 60s, all older than I. As leader of the Farming Systems-Soils (FS-S) project, an older, Japanese-American member built on our discussions to determine the makeup of an ideal team for work in West Sumatra, including one role all agreed I was well qualified to perform.[8] Relations among us were cooperative, curious and intellectually stimulating for all.

His (and the university's) expressed concerns about justice and rationality required an impartial and broad-based search for the 'best candidate'. We all thought I was well qualified (indeed, the terms of reference had been written with me in mind), but he was meticulous about waiting through a long search process reinforced by university requirements, thus contributing to equity, but also approval from his employer and thence to his own breadwinning commitments.

During this process, I also experienced numerous occasions when my colleagues (men) exhibited caring behaviour toward me: For instance, one older married colleague at a nearby centre (who had also unsuccessfully proposed a sexual relationship) expressed his dismay when he learned that my two children and I lived in a cement block apartment house under the freeway near the university—all I could afford on a single person's salary in Hawaii. The FS-S project leader initiated a process that resulted in a promotion from assistant specialist to associate researcher, an added financial burden for his project—partly stimulated, I believe, by his concerns about my inadequate income.

Americans interacting in field and forest

Leaving Hawaii to go to Sitiung, I worked initially with the team described in Chapter 5. My colleagues were intrigued by both FSR&D and anthropology and recognized me as usefully expert in those fields. Although not used to working with women, they were open and accepting about it.

Connell (1995) describes the masculinities he encountered among men involved in the environmental movement, which applied to most of the local FS-S team. The former accepted

> *A practice and ideology of equality.* The common sense of the [environmental] movement includes these principles: no one is supposed to be boss; workplaces are run democratically; no group has rights over others;

> decisions are made by consensus. There is a sharp critique of hierarchy and authoritarianism.
>
> *(p. 127)*

Although one graduate student (who arrived later) came from a fundamentalist missionary background and kept tight control over his wife and children, this did not fit well with the team 'culture'.[9] He was, however, able to deal equitably with me—perhaps because I was older, accepted by more senior team members and had my PhD. In general, the researchers tried to follow the principles Connell identified. Great efforts were made to include spouses, including my own, in our work, to take advantage of their skills, with recognition that women and men both had valuable contributions to make. The fact that only the men (excluding me) had 'real jobs' reinforced the breadwinner harp string that these men (and their wives) accepted and valued; and structurally strengthened a view of women as 'support'—regardless of the principles we all espoused and tried to live.

The team leader, in his mid-60s, was waiting for retirement, and hoping to increase his retirement pay, which would be increased by the higher salary offered in Sumatra, vis-à-vis that of an Island University professor in Honolulu. He also hoped to transfer some of his extensive knowledge.[10] He expressed his masculinity primarily by a fascination with machinery, involvement in the politics of project management (including frequent trips to Jakarta and its attendant luxuries) and caring for the team. His wife was a homemaker, and he had the most conventional attitudes about women within the team (attitudes his wife appeared to share, usually accepting his dominance and control with equanimity).

Although we generally got along well, on 9 July 198X, I complained about him in my journal:

> [He] stayed with me til the end, reminding me unnecessarily about my magazine every time I picked up my purse. His thoroughness can be handy . . . and also an aggravation. He acts as though I'm unable to do anything on my own, despite his need to rely on me for any Indonesian transactions. I know I shouldn't feel this way as he *is* trying to be helpful . . . It's just that I sense an attitude that underneath it all, somewhere I *must* be helpless or forgetful or somehow incompetent to manage on my own.[11]

Later this attitude was again reiterated. The Honolulu-based team decided that the team leader needed to be replaced in July 198X and, to my surprise, asked if I would take over (something I already did informally for his frequent trips to Jakarta). I agreed but specified that they needed to tell him right away, as I didn't want to be working with him under false pretenses. Despite agreeing to do so, by September they had still not acted.[12] I wound up telling him myself, tears in my eyes, rather than continue with the dishonest fiction that all was well.

> He interpreted my tears to mean I didn't want to be team leader! . . . [He] asked [the Southern State soil scientist] to be team leader while he was gone, because [he thought that] I was so upset about it! It all gives me a headache.
> *(journal, 18 September 198X)*

His interpretation that I was upset about being team leader is additional evidence of both his caring and protectiveness—as well, perhaps, of a narrative of women's inferiority. The comforts of life in Jakarta were a draw to him, but they were also the best site for engaging in the political manoeuvring required for obtaining machinery and equipment for the project—all common masculine harp strings. Engaging with USAID, our donor, and to a lesser extent the Indonesian scientists at our partner organization on Java, allowed him to function in English; and functioning in English allowed him to exercise articulateness and sophisticated speech, something that was impossible in the field, where the sounds of Minang, Javanese and Sundanese surrounded him.

Interestingly, all these American men working in Sumatra were of a religious bent, an orientation often assumed to accompany sexist ideas. Yet their behaviour was among the least sexist I have encountered anywhere.[13] Those in Honolulu worked for an American university, as hierarchical a structure as one can imagine. Yet all locally professed and tried to act out egalitarianism.

Americans and Indonesians interacting in field and forest

These egalitarian impulses, so strong among the Americans, came up against a very different world as we interacted with our Indonesian partners. The Indonesian academic world was dominated at that time by strongly hierarchical Javanese and Sundanese men from the Soils Center on Java. All our partners initially were Javanese men, led by a Javanese man with protective and paternalistic attitudes toward his team, which was reciprocated with extreme respect, loyalty and obedience toward him. Most team members had only bachelor's degrees from Indonesian universities, which at that time were 'bootstrap organizations'—new graduates from college taught classes, with very few advanced degree holders (or books) available. We always initially had one Indonesian scientist with a PhD from the West connected to the project, but that man was almost never on site—being drawn away by his superiors who required him for other projects. He was also drawn away by his commitments to his family, living on Java. Besides missing their families and homes, these Indonesian men saw being in the field as detrimental in terms of office politics: they feared 'out of sight, out of mind'.

Another difficulty in implementing the egalitarian ideals of the American team members was the dramatically inequitable access to resources.[14] Funded externally, the Americans each had new jeeps with drivers available to them, with adequate funding for fuel; the Indonesians had one ancient Landrover for all 15 team members, with themselves as drivers, and very limited funds available for fuelling and repair of their old gas-guzzler. Each expat worker had his/her own quite adequate house provided; the Indonesian team members were initially all housed in one tiny concrete block room, sleeping on mats on the floor for over a year, until a larger facility could be built for them. Expats brought their families with them; Indonesian staff came alone. Salaries for expats were at least ten times those of Indonesian staff members (enforced by the government of Indonesia to prevent low-ranking employees from getting salaries greater than high-ranking bosses). Efforts to implement an egalitarian workplace were seriously hampered by these harsh realities. The American team shared transport, inputs and field expenses, but Indonesian staff remained disadvantaged vis-à-vis Americans in terms of mobility, scientific knowledge, communication networks, resulting ability to act on their own initiative and independence. These disadvantages, combined with hierarchic world views, made real equity between the two teams impossible. Insofar as salaries were seen to represent breadwinning capabilities, the Americans' (unearned and inequitable) comparative success probably felt like failure to some particularly competitive partners, though most remained remarkably collaborative (aided perhaps by their sense that social hierarchies were 'natural').

A recurring disagreement between the international FS-S project and the Indonesian partners on Java related to their respective contributions. USAID and the US-based American men running the project complained that Indonesia was not holding up its part of the bargain, which specified equal financial contributions. The emphasis on these funds reflects another harp string of great masculine importance: a love of *quantification* (see e.g., Clarke and Hamilton 2013). The US-based Americans evaluated quantitatively and financially, and Indonesian field budgets were miniscule. Because Indonesian salaries were so low, even though the Indonesians were providing *three times* the number of people the Americans were providing, the total Indonesian budget, and thus externally perceived financial contribution, was far lower.

The comparatively minimal Indonesian gender differentiation may also have contributed (subconsciously) to the American team's acceptance of my own abilities. On the other hand, the perceived lack of masculinity (by American standards) of Indonesian scientists may have contributed to less respect for them.

The harp frame among Sitiung's researchers

Thinking in terms of the three sides of the harp's frame, the broader context in which these actions and beliefs played out, there was again a recurring balancing of *competition* and *cooperation*. At a macro level, there was competition between the two universities involved in the FS-S project. Although an element of this competition was based on *rationality* (concern for obtaining and maintaining funding, important also for *protecting* project employees), part was also based on an acrimonious disagreement between two intelligent, articulate, highly accomplished and well-respected men with significant power, *competing* within the world of soil science, for their respective universities (the 'combatants' discussed later).[15] This competition had the potential—a potential we were able largely to avoid—to affect adversely our work in the field, which was remarkably *cooperative*.

In related fashion, a balance was sought between *hierarchy* and *democracy*. The structure of the overall project (which included work in other countries), was hierarchical. A programme leader at Southern State was superordinate to the two scientist 'combatants' just mentioned, who in turn supervised those of us in the field (three from Island University, two from Southern State).[16] Island U.'s scientific 'combatant' was responsible for the Indonesia site, so had some authority in theory over the on-site underlings of the Southern State 'combatant'. Southern State's 'combatant' was in charge of a comparable site in Latin America. Our site team's commitment to a democratic approach to decision-making and action (despite the existence of a team leader who *could have* approached decision-making autocratically), combined with geographical distance from these superordinate decision-makers, meant that we had significant autonomy to ignore the distant tensions and structure our interactions (at least among the Americans) in egalitarian ways.

Broader religious concerns also played a part, in the third side of the harp frame, in two ways. Locally, two of the particularly religious team members held opposite views on the *controlling–protecting* differentiation. One emphasized Christian religious injunctions to 'love they neighbour', to protect the weak, and to experience and spread joy. The other emphasized the Christian religious tenets that a man should control his household as God controlled the world, and that straying into sin was an ever-present danger. These differing orientations had implications for the ways their households were run, their perceptions of *fatherhood* (important to both) and also tied back to the *hierarchy* vs. *cooperation* issue, on a micro-scale.

At a broader scale, ultimately a perception of *competition* from Muslim men (local, regional and national) about *Islam vs. Christianity* resulted in the expulsion of the man with a more loving, protecting orientation (see Box 5.2). His genuine generosity (*cooperative* spirit) was perceived by some local Muslims as an attempt to 'buy' villagers' faith, and thereby compete with Islam—a perception that made its way up through the Indonesian governmental bureaucracy, eventually reaching the American embassy (as well as long-enduring misperceptions within the international soil science world). In Indonesia, religion—then and now—has been closely linked with politics, both organized by men.

Men in a Dutch conservation organization and a Scottish consulting firm[17]

In Danau Sentarum, West Kalimantan (1992–1993), a seasonally flooded and lowland tropical forest conservation area, the on-site team included my husband and myself, for most of the year with only our boat drivers for partners.[18] The project began because of the aforementioned 'will to improve' of a man who worked for a conservation non-governmental organization (NGO), Swamp, and who had worked previously in the area. On this project too, the main decision-makers were men.

Swamp's Bogor office was perpetually short of funds and had a laissez-faire attitude toward financial management. The Danau Sentarum work was sub-contracted to them, under a British-funded project to manage Indonesian forests, via a commercial consulting firm ('the firm'). The firm was conservative in its orientation and used to rigid financial documentation—the firm leader linked his work tightly to the quantitative harp string, focused on financial management. Again, we see two world views (as well as masculine identities) collide, far from the local research site.

Clothing represents a useful symbol of this division, as it did in Bushler Bay:[19]

> The [Firm] Project Coordinator, Jives, was suave and sophisticated, perfectly groomed, impeccably dressed, and oozing charm in urbane, social and bureaucratic contexts. . . . [O]ur project officer, at [Swamp], was an ex-hippy who had abandoned his earring some years back under threat of expulsion from Indonesia, but had otherwise retained much of that demeanour. He was comfortable with fieldwork, science, and very casual attire.
>
> *(Colfer 2006, p. 4)*

Elite Indonesian men, such as the head of the agency managing parks in Pontianak, dressed smartly, with strong disapproval of sandals and shorts, despite the tropical clime. This was reinforced to our dismay when a British NGO volunteer whose knowledge and labour we valued highly was summarily expelled from the country—his informal attire considered mainly to blame.

The difference in clothing between the two conservation project leaders captures within it, the values differentiation Connell (1995) identifies between corporate and scientific management—one that plagued this work:

> Historically there has been an important division between forms of masculinity organized around direct domination (e.g., corporate management, military command) and forms organized around technical knowledge (e.g., professions, science).
>
> *(p. 165)*

In terms of harp strings, the consulting firm project leader was into control and quantification[20] and the NGO leader was into science and the protection of nature. They did not mesh well.

My husband and I were both hired and wanted to share team leadership, fearing marital discord if one supervised the other. Swamp did not like the idea, despite our both having had team leadership experience; they feared a lack of accountability. I was surprised to see this commitment to hierarchy in an organization that seemed to accept many of the egalitarian ideals Connell identified above. I wondered, for instance,

> Why were they unwilling to explain their rationale in determining our salaries? Was the heavy-handed, authoritarian tone to their faxes something that would continue throughout the project?
>
> *(Colfer 2006, p. 5)*

Other masculinity-related issues bedevilled our work, including debilitating competition between the two coordinating institutions, an enduring lack of faith in social science and qualitative approaches and an utter lack of responsiveness, caring, about field needs (Colfer 2006, p. 34). This may have been related to a perception, at least in the consulting firm, that, as workers in the field, we represented the lowest rung of the prestige ladder. The absence of anyone plucking the caring or protective harp strings for people was a surprise, given my prior experience; Swamp personnel did express such feelings for nature.

Technically formal forest management in this site included the Natural Resources Office of the National Park Service, though the absence of such personnel has already been mentioned. The central government had also given a concession to the Indonesian military (Jamaker), designed to (a) ensure their presence to protect the adjacent border with Malaysia and (b) contribute to funding normal military operations through timber sales. No one knew, however, where the protection forest ended and the production forest began. Logging was conducted via contracts with local loggers, even less transparently than in East Kalimantan (Chapter 4).

Men in the Forestry Research Institute of Indonesia and beyond (1994–2009)

Turning to my experience with the Forestry Research Institute (FRI), I found, as did Killick (1995), an expat in Korea, that "a surprisingly coherent culture has been created by people of diverse backgrounds whose only common bond is not being Korean" (p. 88). My co-workers at FRI had more in common than simply not being Indonesian—we were all interested in forests and most of us were also interested in people—but we came from diverse backgrounds, ethnicities, religions and nationalities, as well as genders and ages. Below, I turn to the masculinities I encountered every day while employed by FRI at their headquarters in Indonesia.

The office was in an experimental research forest managed by the Indonesian forestry research agency. My professional co-workers, for the most part, had PhDs, in a variety of fields. The central mission of FRI was to improve forest management in ways that benefitted both the forests and the people residing therein. FRI was unusual in its acceptance and involvement of social scientists as comparative equals in the research agenda when I worked there full time. In 1994, when I served as a consultant, there was one full-time woman scientist (also American). By 1997, FRI was singled out as doing an exceptional job, compared to other international centres, of hiring women as international hires at 25% (Merrill-Sands 1997). In those days, a couple of members of the management group were usually women,[21] and one director general was a woman (200X–201X). I was a programme leader (an intermediate position) for several years and was also the staff rep on the management group for eight years. Most of those providing administrative and secretarial backstopping were Indonesian women (including a number of Chinese Indonesians). Drivers were all Indonesian men, most Sundanese; and there were mainly Sundanese men as office assistants.[22]

Ethnically there was always a preponderance of white men from the global North in charge (as is common in international organizations), with higher percentages of professional women from such countries as well. However, by the early 2000s, a few Indonesians (men and women) were hired as 'international staff' (a designation that included much more generous salary and benefits than 'national staff' received).[23] Throughout my life at headquarters, there were serious efforts to recruit professionals from all over the world, with varying degrees of success. Recruiting women professionals from the South proved the most difficult. Experiences recounted later in the chapter may shed light on this difficulty.

Masculine harp strings and forest professionals

Forestry is typically portrayed as a masculine discipline; dealing with forests is seen as a task for men. Traditionally, men work as loggers and foresters, not women. The field has had a para-military tradition, including wearing uniforms, handling heavy equipment, shooting poachers and growing, protecting and felling big trees. The necessity to fight forest fires in many regions has emphasized military-style chains of command, *esprit de corps* and dangerous physical work—all associated with masculinity, at least in the global North (see e.g., Eriksen 2013, on Australia and the US).[24] This tradition had begun in Europe, specifically Germany where scientific forestry began, and was taken up by many former European colonies, including the US and Indonesia.

Whereas this was a tradition with which my colleagues were familiar, it was not the gist of their own daily lives. FRI researchers were basically academics, *studying* forests (and forest peoples) rather than *managing* them—many had however been exposed to forestry's hyper-masculine context and narrative for several years in grad school and/or their home cultures, and all were aware of it. We confronted more intact versions of this view among our partners in many countries of the South. As worldly and educated men, however, FRI staff had also been exposed to ideals of gender equity. We were under pressure from donors, our Board of Trustees, and others, to increase the number of women and researchers from the South. Most tried to comply with, in some cases actively support, efforts to address women's professional disadvantages. Indeed, Anderson (2009) concluded, based on his study of university men in the US and UK, that "in the process of inclusive masculinities proliferating, gender itself, as a constructed binary of opposites, may be somewhat eroding" (p. 155);[25] this may have been the case at FRI, as a work context, as well.

Similar to the Kenyah in muting gender differences in office life, there was one important difference: Most FRI staff came from national and sometimes personal traditions marked by ideologies of female inferiority, some with strict, somewhat oppressive gender roles. We all had to fight such underlying, often subconscious assumptions and tendencies.

In the next sections, I identify the relevant harp strings operative in this context that was marked by such cultural and ethnic diversity, by its definition as a breadwinning site and with a topic widely considered masculine as its central mandate.

Men as breadwinners

We were all workers at FRI or partners. The assumption that men were the breadwinners was evident early on. The subsequent discussions in this chapter all relate to masculinities in the breadwinning context. Here I only highlight evidence pertaining to gender dynamics in this predominantly masculine setting and the symbolic significance of attire.

Before I was hired in the 1990s, I was discussing the terms of possible employment with one of the men most interested in hiring me, a white economist of middle age, while walking down the stairs at FRI headquarters. As I recorded in my journal,

> I was right to be skeptical. [Another senior European administrator] asked me two questions, 'What's your husband doing?' and 'When's his contract up?' 'Good bye.'
>
> When I asked [the Australian man] my rank, [he] said, 'Does it matter?' When I asked the salary, he expressed astonishment . . .
>
> *(4 October 199X)*

Why would I care? He clearly assumed that my gender, my obvious motivation to work and the probability that my husband's income would be adequate, meant he needn't worry about wage or benefit equity. The same day, I was also told by other personnel that the senior administrator did not like anthropologists, women researchers or Americans—a damning combination in my case. A few months later, another American woman social scientist applying for a job had a similar experience (journal, 19 August 199X). Later still, a French woman anthropologist whose methods were more typically qualitative and ethnographic than FRI work in general gave a talk in which she stressed the symbolic significance of benzoin trees in North Sumatra (where the milky exudate is described locally as mother's milk). The most quantitatively oriented staff members rejected her talk and her work, though legitimate from an anthropological perspective, treating both with disdain.

After a year there as a consultant, a longer-term job was advertised. I learned some time later that whereas the search committee had selected me as the best candidate, the senior administrator had hired instead three white men from the global North. Although eventually I was accepted as a legitimate member of staff, it took the prime decision-maker a couple of years that I found painful to decide I was capable.

As was the case of women in Bushler Bay's school, I recall much more public discussion and negotiation involved in hiring women than in hiring men, which may be partially, though not fully, responsible for the small number of women FRI managed to hire early on. Injustices occurred related to the hiring process for men as well, which seemed from my status as a researcher, to represent evidence of a harp chord composed of control, rationality and a lack of empathy (related to Northern notions of men's avoidance of emotionality).

The significance of attire as an indicator of work status and values has already been discussed (Chapter 3). It was less informative at FRI than in some locales, though it could affect employment. Northern men, and thus FRI's leaders, tended to wear slacks and short-sleeved shirts, rather than the business suits described by Brandth and Haugen (2000). One from Germany and another from South Africa wore colourful shirts outside their slacks, sandals and scruffy beards. These were not appreciated by the Indonesian bureaucrats we sometimes visited in their offices. Seen as a mark of disrespect, on occasion this demonstrably reduced Indonesian willingness to collaborate. We found suited elegance, à la France, among the African men we worked with as partners (academics and government officials) from Central and West Africa.

Rationality harp string

There is a common linking of rationality with masculinity,[26] with a long history, reinforced by religion, formal education and many government policies (also seen in Indonesia, Chapter 5). This link has been further reinforced by much spurious and previously mentioned 'science'. At FRI, being rational was often counterposed to being emotional, with the latter avoided. Northern men were taught, rather harshly, never to cry. I have never seen a man cry in a work setting, no matter how distressed he was.[27] Even remembering witnessing the genocide in Rwanda or a wife's infidelities, my colleagues were able to hold back their tears. When a woman scientist cried while complaining to a senior administrator about a negative performance appraisal, colleagues (and I) considered it a manipulative move. Northern men were expected to—and did—take pity on crying women (perhaps one reason some men prefer not to hire women). Allowing more emotional expression to both men and women would make the workplace more inclusive and more emotionally healthy.

Connell (1995) lays out a related dilemma: Rationality is supposedly for men, but rationality dictates making use of women's observable talents. Subsequent experience of rationality in women[28] weakens the narrative that rationality is only for men. The very quality that can oppress, rationality, can also be used to contribute to treating women as equals. It includes in its nature its own demise (or at least weakening). However, in the process I suggest, it also strengthens the value of rationality vis-à-vis emotions, ethics and other values. This may in fact be counterproductive in the long run, as a more humane world would acknowledge emotions and other qualities deemed by many to be womanly ('the feminine') more fully.[29]

FRI, as a research institute, put very high value on rationality, as did its research staff. This was evident in discussions, division of labour, performance appraisals, promotions and the regard of one's colleagues and supervisors. It is unlikely that any FRI staff member would have denied that rationality was something he (or she) strove for. It was necessary to maintain one's breadwinning role in this case. And it was also a key consideration in researchers' antagonism to hierarchy[30] (discussed later). Hierarchy was seen as involving an often-illegitimate use of power (which was widely felt 'should' be based on rationality, which in turn would grant more legitimate authority).

The language harp string

Professional men were expected to be able to speak publicly and clearly. At FRI researchers appreciated, in themselves and others, the ability to speak articulately, rationally, concisely and in an interesting manner (see Box 6.1).

BOX 6.1 ANXIETIES ABOUT PUBLIC SPEAKING

I was standing in a garden with three professional men, during a 'get-to-know-you' wine bar at one of the hundreds of conferences and meetings I attended in a country I forget. One man from South America, with another international organization, was pontificating about how he was never nervous about giving a speech. He went on at some length about how no one who was competent at his profession would feel any anxiety about giving talks; the other men nodded appreciatively. One or two expressed their agreement. I suggested that some people were shy about speaking publicly, and I didn't think it necessarily had anything to do with competence; I suffered myself from some such anxieties. The pontificator completely rejected this notion, insisting that if one were capable, one would have the self-confidence needed to avoid all such anxiety.

I have sometimes experienced what some have termed 'mansplaining' and other ways of dismissing women's input, e.g. a case in Cameroon:

> Gave my talk this AM—which generated a lot of discussion. Even though the chair convened the session 18 minutes late, talked for another 8 minutes

> himself and then tried to cut me off about 5 minutes into my 15-minute speech, which was intended to leave 15 minutes for discussion!
>
> *(journal, 15 November 199X)*

Since the language specified for FRI work was English,[31] this gave native English speakers a considerable edge in the office (though not necessarily in the field). One young but capable Northern researcher struggled to overcome his inherent shyness and suffered horrid anxieties whenever he was called upon to speak up in a group (an issue seen as more common among women than men). He saw this as a major impediment to his professional advance.

An Australian researcher gave us all a lecture on good form in making presentations, and we were routinely called upon to speak publicly in many different fora. The fact that we worked in many linguistic contexts meant that we also depended on each other for help when called upon to give presentations in languages in which we were not necessarily fluent (see the section on the caring harp string). We were also expected to write articulately, logically, concisely and in an interesting manner in what we produced—qualities men (and I) also sought in ourselves.

Love of math harp string

In Chapter 3, I mentioned the links among math, science, prestige and masculinity within Bushler Bay's school, something that emerged again in other projects as a focus on finance. Americans of my age grew up thinking math was the quintessentially masculine subject—we were taught this in school via various mechanisms, whether role models or explicit statements about masculine abilities vis-à-vis feminine shortcomings.[32] I remember being told by a teacher, a man, in high school, when I enrolled in a physics class, that this was a 'hard class', that girls weren't any good at it and that we girls would be better advised to drop it (which I foolishly did). Having heard such stories and advice throughout my childhood (my mother was dreadful at math), I grew up with a math phobia; I was only able to address it when I returned to graduate school in public health at the age of 35, determined specifically to overcome that fear.[33]

Such beliefs and assumptions, expanded to quantitative methods and approaches generally, also pervaded the global scientific community. This pattern was particularly evident in the US, a country with disproportionate influence on the global stage. The US provided international assistance; it had a strong voice among multi-lateral donors, and within international organizations, whether research or implementation, whether industry or NGO. Americans' gender biases came along with us.

The most insistent about encouraging mathematical/quantitative interpretations and approaches at FRI have been 'Northern' men (who also tended to have higher positions within the organization).[34] These concerns in many cases represented genuine beliefs that quantification spelled science, our *raison d'être*;[35] but in other cases, individuals convinced of the importance of qualitative methods continued to require quantification because of their perception that evaluations of FRI research would plunge if qualitative work were emphasized and accepted.

I see math, science and manliness as intimately connected at FRI and beyond.[36] If we add to this the fact that academically inclined men may *not* fulfill many of their own cultural 'ideals' of masculinity (e.g., physical strength, work outdoors, sports), a focus on math may be a sensible harp string to pluck.

How has this played out in practice? Around 2000, several of my colleagues and I decided to develop a research programme on adaptive collaborative management (ACM, Colfer 2005). By this we referred to attempts to build on the kind of collaborative work discussed in Chapter 5—working with forest peoples—but also incorporating more significant ecological components than we had in Sitiung. Although ACM included the probability of doing some quantitative analyses, the fundamental approach was qualitative, participatory action research. We were convinced that to change local systems such that management could be truly sustainable—as was FRI's mandate—we would have to involve local people in a serious way.

Unlike other teams doing more conventional, quantitative studies, we had to hone our proposal through *17* iterations, and I (as team leader) had to defend it and re-defend it at every meeting of our Board of Trustees for the first three years (Colfer 2013). That I was able to do so successfully each time suggests that our plans and implementation were solid. Senior administrators' initial discomfort with the approach was obvious,[37] despite broad recognition that our arguments in favour were compelling. Several subsequent research efforts built on this approach, including some by these administrators.

Masculinity and control harp strings

Some researchers emphasized and studied forests primarily for timber management; some for conservation; and some for plantation management. The common masculine harp string of control was differentially selected by these FRI staff. The desire to manage is a kind of control, with those concerned with plantation management the most interested in this harp string. For such researchers, control was closely linked with production, another idea affiliated with masculinity. The idea that production is to men as reproduction is to women is hegemonic among many groups. Production is linked as well to the provisioning harp string. There's a masculine chord—dominant on the global stage—that links control, production and provisioning (or being a breadwinner for one's family).

At FRI, plantation management was never dominant. FRI's umbrella organization had the formal mandate to increase production, protect the environment and focus on 'the poor' globally. But the group who created FRI, including all of its early Directors General (DG), wanted an institution that addressed policy and people issues more than production per se (the latter more common in most related centres). These institutional concerns strengthened FRI attention to more 'natural' forests, specifically those that were being managed for timber but might be managed in a more holistic fashion, and those managed for conservation.

Those researchers interested in conventional forest management were trained to manage for timber but were trying to understand and incorporate findings related to other forest uses, by the variety of forest users (differing by gender, ethnicity, class, etc.), and policies that might enhance forest sustainability. Such efforts also involved masculine control, but of a more open-ended variety than plantation management or agriculture.

One would think the control harp string would be least prominent for those interested in conservation management, where wildness was so valued. But the strength of their protection (of nature) harp string reinforced the concern for control (as shown also in the Swamp example earlier).

After a long period of considerable cooperation among staff, a new administration believed that excellence emerged from competition, and so competition began to be expressly encouraged among us. I do not believe this improved the quality of our work, and it certainly reduced the level of cooperation within the institution. Yet another administration strengthened control and hierarchy within the organization, firing or failing to renew the contracts of some people with valuable qualitative skills (mentoring, participatory action research, networking) who were less good at meeting deadlines or quantifying research results. This increased fear, which served to reduce the amount of selfless caring behaviour.

In the work setting, the internal conflict between the need or tendency to exert control and men's own resistance to hierarchy also created difficulties for them. The reluctance of the FS-S team member mentioned earlier to inform the Sitiung team leader that he was being fired was one example. Another was the assignment by a senior administrator to one of the most junior people on staff then, the only woman, the responsibility to withdraw a FRI contract the equity of which I had diplomatically questioned (journal, 13 November 199X).

Love of nature harp string

The love of nature and the outdoors were important values for most of my colleagues. Like the men of Bushler Bay, many (particularly those from the North) liked being in the woods and were proud of their ability to endure difficult conditions, hike long distances and withstand harsh temperatures. I remember the admiration my colleagues expressed around 1997, when a man in his 60s, a forester, was able to keep up, trekking all day long through rice fields and forests in East Kalimantan.

And I recall my own dismay when, in my mid-60s (2009), I found myself unable to keep up with my younger colleagues in Madagascar (all men) as we traversed steep terrain. I was aware that my pride in such abilities was related to my long-felt need to 'measure up' in a way deemed masculine by my peers. I, like many professional women, had cultivated such abilities partly in order to survive in contexts defined as masculine, where I chose to work (as noted also by Eriksen 2013 and Reed 2003 for forestry women in the US, Australia and Canada).

Thirst for adventure harp string

The thirst for adventure—which I shared—is also more closely linked with men and ideas about masculinity than with women; it was a common interest at FRI (as among the Kenyah). Our jobs took us to many countries, mainly in the tropics. Exploring new places, experiencing alien conditions, meeting strange new people—these were common values among my colleagues. I remember the thrill of my first (1979) experience of travelling up a small stream in a canoe in Borneo, with foliage from each side meeting overhead. Or the excitement of recognizing similarities between the villages I was seeing along a river near Yurimaguas in Peru in 1983 and communities I'd studied in Kalimantan. Or how intrigued I was, surrounded by black people in Côte d'Ivoire on my first visit to Africa in 1995. These experiences were utter delights, and I saw the same excitement and thrill in the faces and words of the men I worked with as they visited new places and met new peoples. One of my research partners, an American academic, wrote a fictionalized version of his own life in a tropical country, which nicely captures this common masculine harp string (and others; Moro 2018). For anthropologists, this is often discussed with disapproval, as the lure of 'the other'. For biophysical scientists, curiosity about nature is more emphasized—but all these adventures include feelings of newness, of potential danger or discomfort, of challenge, of learning, of excitement. It's part of discovery.

The caring and protection harp strings

I described previously the uncaring attitudes and behaviour of project leaders in the Swamp project. But FRI's men were far more cooperative and helpful to each other (and to me). Caring was not so often discussed, in the way that quantification or rationality were, but it was evident in men's behaviour.[38]

There was routine sharing of expertise among researchers. As noted, we often had to give presentations in languages other than our own. French speakers helped English speakers craft better talks in French (journal, 25 June 199X); sometimes one staff member would translate for another; colleagues would correct the English or French or Spanish or Indonesian of a colleague in written materials as well. We also routinely critiqued each other's papers, helping to improve them (e.g., journal, 7 July 199X).

Although the level of inter-programme cooperation varied over time, there was a five-year period when I was a programme leader, when we all willingly shared our budgets, allocating excess to those who had shortfalls. Overall the shared interest in better care of the forests and greater benefit to forest communities tied the researchers together with a common goal and strengthened caring behaviour.

Protection and caring share some elements, though protection can carry with it an element of control (as in the Balinese husband described earlier). FRI policies included institutional protections for staff members; once when I appeared to have disappeared in or on my way to Turkey, great efforts were put forth by my colleagues to find me. Additionally, when FRI staff members of either gender were robbed, as happened several times, long discussions ensued on ways to provide better protection or to avoid further exposure. When Jakarta erupted in 1997 and 1998 into widespread violence, administrators did everything in their power to protect all the staff (particularly the Chinese who had been so targeted). 'Protection' was not, however, used as an excuse to prevent women from performing our work (unlike in some places).

Sexuality and violence harp strings

Sexuality was not an overt part of office life at FRI. I was the staff member who dealt with anonymous complaints about sexual harassment during my tenure at headquarters. Such complaints were rare, and FRI administrators dealt promptly with them when so informed. Warnings were usually sufficient. No woman who complained was willing to make a formal complaint or come forward publicly for fear of reprisals both within FRI and more broadly (29 August 199X). In only one case was disciplinary action necessary (e.g., journal, 9 March 199X).

Such behaviour was said to be far more common within the Indonesian bureaucracy, and the most substantive complaints I heard were from Indonesian women in service positions regarding Indonesian men of high rank. Such harassment was surely more widespread than what came to my attention. One woman was subjected to repeated requests for sex and inappropriate touching; the other was locked in a 'superior's' office and kissed before she was able to escape. They felt both embarrassment and fear for their continued and future employment, because of the harassers' power within local institutions and the acceptability of such behaviour at that time within Indonesian bureaucracies.

Interestingly, I heard as many complaints from men as women. One Northern man complained about harassment from a Southern woman who wanted to make her husband jealous; and one gay man complained about harassment from another gay man, both from the South—neither involved physical actions. The men's fears were of a different kind than the women's. Women feared attacks on their persons. One man's fears were about his reputation; the other's more irritation than fear, though potential public exposure of his homosexuality probably also played a role—despite FRI's increasingly open policies about LGBTQ issues (including hiring at least one openly gay man and having a gay woman on their board at one point).[39]

Shortly after I joined FRI, the DG sent the other woman researcher and me on a field trip to Kalimantan, one purpose of which was to develop public relations (PR) materials that would show FRI as a gender-equitable institution. The PR consultants were particularly interested in the prostitution that flourished around timber base camps, so my colleague and I were filmed interviewing the prostitutes near the timber concession (Chapter 4). We incorporated what we learned in these interviews in a small box in a thick report that we generated a year or so later.

One evening, I was called about 10 PM and asked (by a Ministry of Forestry official, a man) if I were "encouraging prostitution in Kalimantan". It evolved that another man (Indonesian, researcher, forester) working in another country at an international forestry institution had read the small box in our report, photocopied that page, and distributed it throughout the Indonesian Ministry of Forestry with an accompanying letter of complaint against FRI. This struck me as the height of hypocrisy, given the informal support by government personnel for prostitution near timber camps and military establishments, and the 'open secret' about the frequenting of such establishments by ministry officials themselves. FRI's DG's response, thankfully, was to defend our report and our responsibility to publish the findings from our work, wherever they might lead.[40] The ministry eventually acquiesced.

The sexuality harp string tended to be muted at headquarters. Between 1994 and 2009, there were a few cases of marriages falling apart (some involving adultery) and/or re-marriage to a colleague or secretary, but these were rare, compared to other contexts I'd known. People of both genders had spouses who refused to join them, maintaining long-distance marriages; some left because no work was available for their spouses. For many, the choices were abstinence or the nightlife scene; my sense is that more chose the former.

A few (single, to my knowledge) researchers sought sexual release in bars in Jakarta; in our African locations, sexuality was more open. One happily married colleague of non-African ethnicity, a man, was offered a prostitute in West Africa by one of our African partners who had already made use of her services (journal, 23 June 199X). The scientist refused on the explicit grounds of fear of HIV/AIDS. Another African partner in Central Africa was criticized, not for making use of prostitutes' services, but for spending so much time at it—interfering with his work performance. He was explicitly admired by his African colleagues for his sexual prowess. An American, short-term team member on the same project (engaged to be married shortly in the US) availed himself of prostitutes' services and suffered guilt and fear about having done so. One of our woman board members and I were both approached by a considerably younger Cameroonian man, suggesting we invite him into our hotel room, and I spent an unpleasant day in a car with a Cameroonian forestry official who made suggestive remarks all day (journal, 20 October 199X).

One interesting twist for expat life in Indonesia was the fact that the inconvenient question of male control in the household could be partially set aside. Men (and their wives) could avoid confronting the incompatibility of 'modern' gender equity concerns and traditional gender roles. These men's wives did not do much in the way of household labour, as local women and some men were hired to perform such roles (discussed later). Childcare was also simplified, as help was readily and inexpensively available. Relations among spouses were in most cases companionate (rather like the Public Employees discussed in Chapter 3 or my own marriages). The globally conventional male family provider harp string was evident and re-enforced (a) by the patterned but largely unconscious FRI preference for men as employees, (b) the difficulty of finding work for foreign spouses there and (c) Indonesian governmental reluctance to grant work permits to them.

The harp frame—FRI

Whether this analogy can be stretched to apply to a work context like FRI remains a question. But there were issues that transcended masculinities and structured life and debates within the organization (the frame of the harp). I have selected three,[41] to continue with the harp frame analogy: the degree of hierarchy vs. democracy, international vs. national staffing (including ethnicities) and headquarters vs. branch offices. As will become clear, these three themes interconnect as they affect and are affected by individuals.

From the beginning, the narrative we told and heard was that we were a 'flat' organization, democratically run. This was a value to which both the organizational statement and the researchers themselves expressly subscribed and for which researchers fought. But from the very beginning, what was perceived (from below) to be autocratic decision-making on the part of DG occurred. The policy manual, after listing all sorts of democratic sounding procedures to be involved in hiring, for instance, concluded by saying basically that the final decision was up to the DG—a right several DGs regularly exercised. The degree to which a particular programme replicated this kind of control also depended on the personality of the leader. Some kept tight control of budgets, for instance, while others devolved such control to specific project leaders. Some gave explicit orders (with varying success), while others sought agreement from their 'underlings'. There was ongoing oscillation as leaders' felt needs for control varied. But certainly the exercise of control made those with formal authority, particularly those who regularly exercised it, exceedingly unpopular. FRI's researchers 'loved to hate' those in power.

Another factor influencing hierarchy was the growth of the institution. As an institution grows, some bureaucratization becomes inevitable due to the limit on how many people one can deal with meaningfully on a regular basis. A third factor was the amount of funding available. When funds were abundant, there was a loosening of control; when funds became scarce, a tightening occurred. But where we stood on the hierarchy-democracy continuum was a topic for continuing discussion and strong opinions—influenced by and influencing the strings of the harp.

The second side of the frame, the differences among ethnicities and nationalities, was a pervasive structural issue, as noted in footnote 23. At its most macro scale, it played out as the distinction between international and national staff, with the latter singularly disadvantaged in terms of salaries, benefits and prestige (cf. earlier discussion of FS-S project, also replicated in the Swamp project).

Although salaries and benefits were the inequities most typically raised, the prestige factor was probably more debilitating for national staff. The most vivid example in my memory came when a visiting Northern man met with a group of researchers, including a very well educated, articulate and capable Indonesian woman specializing in a technical field. As we sat in the meeting, the man kept turning to me—an anthropologist without significant knowledge of her field—with related questions. He kept this up, despite my repeatedly pointing to her as the expert. It simply 'did not compute' in his mind that an Indonesian woman could have such expertise.

A junior Indonesian man, a forester, led one project on which I also collaborated; when we met with outsiders, all assumed I was the leader, rather than the very capable Indonesian man. These kinds of experiences, repeated regularly at work, were demotivating and emotionally harmful for national staff. Many Indonesian staff indeed did not have the same academic skills that the international staff did, but many did, and many who began with low skill levels improved over time, without adequate acknowledgement from international staff or visitors.

In the battles over hierarchy vs. democracy, it was the international staff, those with greater power within the organization, whose voices were most strident. The national staff, Indonesians at headquarters, had more conflicted feelings about the situation. Their own cultures tended to be hierarchic (as discussed in Chapter 5), yet the crass inequities,[42] obvious on a day-to-day basis, could not help but influence their thinking in this context where rationality and equity were so highly valued.

In Java, power was also very differently manifest than was the case among Americans, Europeans and Australians. Traditionally, a powerful Javanese was quiet, his authority taken for granted, emanating from spiritual strength and the respect of those beneath him in the hierarchy. Persuading Indonesian researchers to speak up, to share their thoughts or even to complain about what troubled them required special effort (which many international staff were unwilling to put forth). Some took the silence as acquiescence or acceptance, some as timidity or incompetence. Adherence to religious dictates was also highly valued among the national staff. The higher an Indonesian's education, the more likely was he to emphasize his religious observances. The reverse was true of those from the global North.

In the Central African branch office, FRI used the office complex that also housed another centre under the same umbrella organization. There, relations (in the 1990s) between international and national staff were far more conflictual. A triumvirate of American and European men ran the office with an iron hand, without apparent attention to equity issues (or common humanity) related to either gender or ethnicity.

> [An American woman scientist] has a rough row to hoe. The only woman on staff not sleeping with one of the other scientists. She starts to talk and they begin reading something in meetings. They've driven away the

> previous two women scientists [confirmed by me]. But she doesn't want to bring it up [with outsiders] for fear of this problem being used as an excuse to cut their funding. They're already barely scraping by.
>
> *(journal, 6 November 199X, also 27 October)*

The FRI DG and other leaders at that time were sufficiently concerned to negotiate a sharing arrangement that kept FRI researchers outside these sexist, racist and authoritarian leaders' control (journal, 28 October 199X).

Later, in the 2000s, FRI's administrative structure took on a more controlling aspect, subscribing to the very American insistence on bureaucratic control and timeliness, both of which conflicted with Central African laissez-faire attitudes, particularly toward punctuality.[43] This resulted in the firing of some excellent people—people, both men and women, whose talents lay in mentoring, networking, policy influence, community action and the like. Some of these talents, though clearly related to FRI's mandate, did not sit well with masculine ideas of a Northern research institution's 'normal' functioning nor did they carry the same prestige weight as writing a paper or publishing a book—including under a woman DG.[44]

There were also inequities that affected the international staff, particularly relating to language. The strong emphasis on being articulate was far easier for those whose native language was English than for those for whom it was a second (or third) language. This could affect researchers' appearance of rationality, their articulateness[45] and their productivity, thus potentially influencing their breadwinning capabilities, as well as their self-confidence as men.

The third frame, headquarters vs. branch offices (or centre vs. periphery), involved structural disadvantages in that the Indonesia office was far better equipped with modern technology, support personnel and other resources than were the branch offices. Yet a central FRI policy was for researchers to operate in interdisciplinary teams and in partnership with researchers in tropical countries. Almost any project involved people at headquarters and people in the branch offices, as well as partners in national institutions like national forest services, universities and NGOs, with even more exaggerated differential access to resources and conveniences.

FRI organizational policy began with the commitment to a 'flat' organization described earlier, but there was a tension and oscillation over time between an ideal of a top-down, centralized institution led from headquarters and a decentralized, bottom-up institution led by national needs and priorities. At most times, an ideal of cooperation among projects and units on a level playing field was espoused.

However, realities made this impossible. Branch offices, of which there were a varying number, operated in languages other than English, which added translation to almost any task they undertook. Although all were located in cities, much work was conducted in forests. Electricity, computers and later the internet were available (in most cases, sporadically) in the office, almost never in the forests. Budgets for branch offices were lower, so access to computers and printers was more difficult. Disease was more rampant, particularly malaria and HIV/AIDS in FRI's African locations, meaning more absenteeism due to illness (of the researcher and his extended family). Whereas FRI had a library at headquarters, branch offices did not (in any meaningful sense), nor did other good local libraries exist to which researchers could turn. Political turmoil was a continuing problem.

These disadvantages of working in a branch office were partly compensated by researchers' interest in being 'close to the action', being closer to forests and outdoor work for some, being able to contribute to science that might be more directly applicable, more 'grounded', for others. But from other standpoints, in competition with other scientists (a common masculine harp string), it was a very uneven playing field.

From the standpoint of breadwinning success, as well as being able to demonstrate such masculine harp strings as productivity, industriousness and articulateness (all valued among this group), these were serious disadvantages. Productivity at FRI meant producing articles, books, workshops and policy briefs.[46] Doing that in a context without reliable electricity or easy access to libraries, computers and printers complicated the process significantly. Distance from headquarters also slowed down any of FRI's bureaucratic or budgetary requirements, as communications were not always quick and headquarters personnel often had other pressing demands on their time. 'The squeaky wheel gets greased', but it was hard to be very squeaky from a branch office.

Finally, there were qualitative regional differences in the forestry world that bear further research. What I can say here is stereotypical but represents broad patterns I observed as I travelled from one forest region to the next. It seemed, for instance, that men working in Latin America[47] were more comfortable with dramatic emotion than were those working in Asia or from the North. Latin American partners readily expressed their likes and affections. They also argued, sometimes passionately, showing anger, conflict, giving free vent to their dislikes . . . then returning to friendly relations—deemed 'too emotional' by many from the North. In Africa,[48] there was a willingness, once rapport was established, to engage in intellectual debate, showing their own rationality, cleverness at argument, linguistic skills and store of knowledge. Strong African commitments to family in a context of much disease (and thus missed work, ignored deadlines) was an issue for those from the North. In Asia,[49] emotions seemed subdued, with efforts made usually to present a respectful, patient, gentle and caring demeanour . . . until some drastic conflict occurred, a tipping point was reached and relations might then be cut utterly and forever. Those from the North and Africa were frustrated by Asian reluctance to engage in debate, to contribute their ideas.

Masculinities in the domestic world

This discussion would not be complete without some attention to the world outside work which these men (and women) inhabited. At work, there was at least a fiction, and sometimes a reality, of equality between men and women—with women tending to fit in with this masculine context.[50] But this was quite distinct from domestic arrangements, where wives, whether working or not, typically remained the household managers, in many cases complicit in encouraging or accepting the control of their husbands,[51] and often managing men from local ethnic groups. At home, at least when husbands were away, the gender hierarchy was partially flipped, with an ethnic hierarchy gaining prominence. The work these wives did managing the household contributed a vital support function that their professional husbands would have had difficulty doing without (organizing meals, hygiene, childcare, etc.). Such support was particularly important because of the heavy travel schedules of FRI staff. At the same time, wives did not tend to have onerous domestic tasks beyond household supervision, as these were performed by men and women from the country of residence. Unemployed husbands were rarer, but most working women continued to do most of the household management, at least when in country. My own husband did not suffer as much as many did by the disregard such men sometimes encountered. Although FRI had policies to help find work for spouses, these were not particularly effective. Being a trailing spouse was not easy for most.

The philosophical quandaries manifest in the harp's frame invaded the home as well as the office, though people's abilities to compartmentalize varied.[52] Some were able to maintain strong gender egalitarian attitudes at work, but still expect their wives to manage all domestic work at home—expectations their wives may or may not have shared. Although most FRI marriages I knew well were companionate in form, others involved stark masculine dominance or major decisions made without the wife's knowledge.[53] Some spouses simply left (as was very common among wives in 1979–1980 among the European expats in East Kalimantan).

In marriages where the wife had been professionally active, both men and women were likely to suffer as these women were limited, by the absence of job opportunities and governmental reluctance to grant work permits, to lives of household management. The only option, short of leaving their husbands or persuading them to quit their jobs, was to create lives as household managers.[54] All unemployed FRI husbands had been professionally active. I know of none who regularly took over household management. Some of these men worked as consultants part time, some left, some simply suffered, while taking up some of the slack at home when their wives travelled.

The inequities described with regard to national staff vis-à-vis international staff were magnified in the home context, where household staff tended to be paid considerably less even than national staff at the office. There were various rationales for paying such low wages. On one side, FRI researchers almost always paid higher wages than the average wages Indonesians paid their household helpers. But on the other, when compared with international staff wages, the differences were shocking.[55]

Nor did most workers have what would have been considered fair working conditions in the North: Drivers often worked from early in the morning until late into the night; cooks were likely to do the same, making early breakfasts and cleaning up after late dinners. There was little awareness of 'workers' rights' or fair employment conditions among Indonesian workers, nor did most FRI staff, to the best of my knowledge, tend to share such ideas or implement such practices with their staff. Taking advantage of such workers was the custom and following suit was the path of least resistance (to our shame).[56]

I do not suggest that answers to these inequities were simple. It is quite true that international workers were likely to have ongoing expenses in their home countries, encouraging wage inequities; they were likely to suffer emotionally by being distant from loved ones and missing out on family obligations and pleasures; domestic tasks took more time in developing countries than in the global North, so there was more *need* for household help; hiring more people provided more employment for the local population; etc. It was equally true that women were more likely than men to have had household management experience and skills (though FRI wives were almost as likely as their husbands to have had professional experience and skills). These structural inequities remained shocking nonetheless—and they intersected and interwove with the strings of the harp, which together created this world in need of change.

Significant effort was put forth to create better 'work-life balance' at FRI, as there was always a recognition that people were working too hard, spending too much time away from home, putting undesirable strain on marriages, parenthood and home life in general. Most men at FRI were fathers and valued being fathers, but the harp string they plucked was being a 'family man' (being a responsible provider, usually not straying sexually, engaging with their children), more often than producing many children as evidence of their virility (as is common among some African and Latin American masculinities).

There was always a frenetic element to time management. Everyone felt overworked; system dynamics models showing how 'burn-out' occurred were circulated. But people were unable to stop over-working. This was partially related to the genuine commitments people felt to what they were trying to do: solve forest management problems, make communities more equitable, influence conservation policies, etc. But intertwined with such commitments were compulsions related to the masculine work environment: Be productive! Work hard! Prove your articulateness and rationality! Don't waste time! These were internal commands more than external ones, though they were reinforced by annual performance appraisals in which all had to list exactly what they had accomplished; a continual reminder requiring both competition with colleagues and cooperation with partners in the work setting.

Conclusions

This chapter analyzes work life in international conservation and development contexts that relate to forests (though I imagine the patterns observed extend beyond forest-related efforts), with an emphasis on masculinities there.[57] I began with my early experiences in which sexuality and caring were dominant themes for the men I worked with (1970s and 1980s). In West Sumatra, we looked at national vs. international staff conditions and values, the significance of quantification to American and hierarchy to Indonesian scientists, the see-saw for American men between competition and hierarchy vis-à-vis cooperation, the differing harp strings of controlling and protecting in the mid-1980s.

In West Kalimantan, we encounter again quantification, control and competition as masculine harp strings; a caring harp string that focused on nature rather than colleagues; and attire as symbolic of differentiating masculinities, in the early 1990s.

We then turn, in the mid-1990s to 2000s, to the most recent observations, at FRI, an institution devoted to understanding and improving forests and the lives of people living in or near them. The majority of professional workers at FRI have always been men, and they have been addressing a context that has been defined as masculine (the forest). The women who have worked at FRI have attempted to fit into this masculine domain, many nibbling away at the edges to insert concerns that might be called feminine.

Meanwhile the world at large has, over time, also been nibbling away at the idea that men and women, masculine and feminine, are polar opposites. In recent years, the number of women who study forestry-related fields has blossomed, sometimes surpassing the number of men enrolled; institutions are becoming more gender-balanced. We are beginning to recognize—as Kenyah Dayaks, for instance, have known for some time—the huge areas of life, thought and biology common to both.[58]

Transformation of the world is always underway. The early emphasis I perceived on sexuality, for instance, was surely not unrelated to my own age and that of my colleagues. At FRI, in more recent times, I as well as my colleagues, were generally older than in the previous experiences described. The aging process is surely an important factor in this shift in emphasis away from sexuality and toward issues like knowledge and articulateness.[59]

All in all, greater acceptance of the legitimacy of plucking harp strings like fatherhood, caring, protectiveness, emotional vulnerability, holistic thinking and others often deemed feminine, can, I believe, open up new understandings of the world around us. It can also help to capture the creativity of those whose voices have been so muffled (women, people from the South, forest peoples, etc.). Part of that muffling derives from some men's (and some women's) desires to prove they are in control (thus, for instance, not *controlled by* women or village people and their wishes); that they are articulate (thus failing to create or seek out opportunities to listen to women or village peoples); and that they are quantitatively skilled (thus failing to comprehend the values, systems, interactions and intersections that characterize life in forests and are more accessible by qualitative means).

Notes

1 See Hancock (1989), where the poem, 'The Development Set' appears (4th unnumbered page). The material in this chapter comes from two main sources: my personal

journals, which I have kept throughout most of my life; and my memories, with some reference to field materials at my home and in Cornell's archives under my name.

2 Feminist standpoint theory, used partially here, "uses women's perspectives to describe men's behavior at work" (McGinley 2004, p. 373); see also Chopra's (2011) similar approach.

3 Dove's (1999) study of planters across Indonesia represents a useful and relevant study. He argues the importance of studying power among the powerful as well as the powerless, as I attempt here.

4 As global awareness of gender dynamics has strengthened, many of the men I've worked with now have greater sensitivity to the gender implications of values, behaviour and institutions than these historical experiences suggest.

5 Tim Babcock, who kindly reviewed an earlier version of this chapter, suggested that, "Commonality of international (western derived) organizational structures, rules and regulations might be behind some of this [commonality]" (personal communication, 2 October 2019). Quite probable.

6 Subsequently taken over by an Indonesian firm with close ties with President Soeharto.

7 Another example represents danger: One of my colleagues sexually molested an 11-year-old girl (May 198X)—still representative of some men's ideas of legitimate sexuality.

8 The FS-S team leader was *not* seeking the qualities that McGinley (2004) attributes to men at work, which she sees as reinforcing ideas about men's superiority over women (see fn 9, p. 42).

9 He was also an avid basketball fan (Southern State)—the sports harp string—and an extremely competitive and agile player in our neighbourhood pick-up games.

10 The motivations of all these men (and myself) were influenced by what Tania Li, from a critical perspective, calls, 'the will to improve' (2007).

11 My journal in July 198X lists a whole string of helpful actions on the part of men from the collaborating institutions (Island University, Southern State University, Cornell University and Soils Center) to support me professionally (offering to hire my husband, invite me to Cornell University, give talks at the World Bank, professional meetings, etc.).

12 The Honolulu office's reluctance to do so surely related to their commitment to egalitarian relations, like Connell's (1995) example of the "librarian . . . [who] is wrestling with how to supervise other staff and reconcile his actual authority with his belief in equality. So far the result is a draw" (p. 174).

13 Pascoe (2007), writing about an American high school, says, "In fact the table at which the Latter Day Saints students convened during lunch was (apart from Gay/Straight Alliance meetings and the drama classroom) the least homophobic and sexist location on campus!" (p. 112), though she goes on to note, "Their respective religions buttressed male power through their teachings such that the interactional accomplishment of masculinity was less central to their identity projects" (p. 113).

14 "Today was an emotional one for me as I struggled with inequity on our team. Unfairness. We met and discussed it" (journal, 28 February 198X).

15 Language, for academic men, *can* represent the weaponry in a narrative of warfare. In a book on anthropologist Derek Freeman and his academic conflicts with Margaret Mead and others, Hempenstall (2017) says, "Language was warfare now, and both sides [Freeman vs. Cote] practiced open aggression" (p. 216). He adds later, "On both sides the language of the debates [between Freeman and a number of prominent anthropologists] was the language of rolling warfare" (pp. 252–253). It can of course also be used for a variety of more benign purposes.

16 I encountered no other women (other than secretaries) employed on the broader project, though we occasionally hired spouses and two other women as short-term consultants. Our Indonesian partners had one woman technician, with ~15 men scientists and technicians.

17 Material for this section comes from Colfer (2006), my journals and memories; see also Yasmi et al. 2007; Wadley et al. 2010a, b; Yasmi and Colfer 2010; Colfer, Wadley, and Widjanarti 1996; Colfer, Wadley, et al. 2001; Colfer, Woelfel, et al. 2001; Wadley and Colfer 2004, for more on this site.

18 Like the Sumatran example, the donors had been promised a greater contribution from the Indonesian side, however, unlike in Sitiung, the governmental agency in Pontianak was unable to persuade any staff to join us in such a remote area for any length of time.

19 Brandth and Haugen (2000) discuss two versions of forest men in Norway:

> The power of the forestry worker is based on bodily strength as required when 'battling' with the natural environment. It is displayed by means of working clothes and machinery [see Chapter 3]. The power of the organisational man is based on control over economic resources as he leads and manages other men's interests, displaying masculinity by means of the power 'look' of business suits, conference tables and rostrums.
>
> *(p. 354)*

20 This reminded me of a job interview I had for a job in Bangladesh, funded by USAID, with a population NGO in the late 1970s: I was told that all my work would be evaluated strictly based on the 'number of family planning acceptors' I recruited and that I could not publish without permission. I withdrew my application.

21 During the 1990s, the management group was composed only of men, and usually 75–80% of the researchers were men. By 2019, the management group was 65% men (though staff of both genders complained in 2018 that the decision-makers were four autocratic men); between 2013 and 2019, the percentage of professional men staff was in the mid- to high 60s (15 May 2019 email from FRI's human resources division).

22 Men (compared to women) currently in FRI 'services' (non-'professional') have averaged 40% for the last seven years (2013–2019) with very little year-to-year variation, which is probably about the same as in earlier days.

23 This inequity, which pervades the world of international organizations, is a complex issue: People from the country where the institution is housed were paid lower salaries. At FRI, they were paid, according to the head of admin, a man, 'what the market would bear', which resulted in grossly inequitable conditions. Within my own programme, for instance, two women with equivalent skills doing the same job received widely divergent salaries and benefits—both unjust and destructive of motivation.

24 See Fowler (2012) for an Indonesian context in which fire management is closely linked with women.

25 Boylan's (2008) quote bears repeating: "I remembered a T-shirt I'd seen someone wearing at a conference once: *There are only two kinds of people: Those who reject the binary, and those who don't*" (p. 258).

26 Indeed, Connell (1995) remarks, speaking about 'the affluent': "Hegemonic masculinity is culturally linked to both authority and rationality, key themes in the legitimation of patriarchy" (p. 90).

27 Kenyah men, on the other hand, did not avoid emotional expression. In my own personal background and in Bushler Bay, men never cried in public and rarely in private. I recall my astonishment when my Kenyah colleague, a local leader, stood up to speak at a community meeting on his return in 1980 after a long absence. He said a few words and then began quietly to cry, tears streaming down his face. He cried for several minutes, as those assembled listened silently and with acceptance. Then he resumed his talk.

28 The aforementioned senior administrator's attitude toward me shifted markedly upwards after I wrote a logical and evidence-based memo analyzing the inequities and practical implications of hiring senior biophysical scientists and only junior social scientists for an organization ostensibly interested in interdisciplinary collaboration and people-oriented research (e.g., journal, 4 September 199X).

29 Hempenstall (2017) discusses Derek Freeman's arguments with various other anthropologists: "But Shore was more sad that Freeman persisted in seeing Shore's position as 'fuzzy or weak and unmanly', as though only Freeman's strongly masculine interpretations told the truth about Samoa" (p. 192).

30 My own antagonism to this is reflected in this quote from my journal:

> The National Working Group on C&I [is] an endurance contest. Mostly powerful men, some with hidden agendas, most not wanting to work. How *do* you work thru the system? The higher up I go, the less I see that I like. But I guess this is not a new discovery, just further confirmation of a suspicion I've had for a long time . . . I do not feel comfortable in the 'halls of power'. I don't like their smell.
>
> *(19 November 199X)*

31 In the early days, a Malaysian researcher, a man, wanted to sanction secretaries who spoke Indonesian when at their leisure, because "This is a scientific institution and English is the language of science". Thankfully most were not so rigid about it.

32 See e.g., Fine (2010) or Saini (2017) for careful analysis of both lay and 'scientific' fallacies related to gender, including men's purported superiority in this field.

33 A recent Twitter feed invited 'Women in Computational Social Sciences' to a workshop in Switzerland, saying, "So far the field of computational social sciences is male dominated and short of female perspectives and insights. . . ." (https://twitter.com/Kressin_L/status/1189147458116427782, viewed on 12 January 2020).

34 One of my early journal entries when I began work at FRI:

> The 'science meeting' met from 2–5:45, and they smeared [a European man, a forester with social science interests] all over the floor, and [behaved] generally as white, Anglo Saxon male know-it-alls who gave no credence to alternative ways of looking at the world.
>
> *(journal, 31 August 199X)*

This was in fact the meeting that prompted the aforementioned memo (footnote 28).

35 I recall recognizing and bemoaning the *impenetrability* of some quantitative analyses—so elegant, so admired by society, their adherents so self-confident; yet forming conclusions so fundamentally wrong, based on erroneous assumptions . . .

36 One reviewer, a man, rejected the idea that math is related to masculinity, linking quantification instead with the need for generalization. However, one can do research (perhaps of a more feminine nature?) the purpose of which is not generalization, but rather understanding of particular contexts, both for practical use in that context, and also for insights about possible interactions and phenomena elsewhere.

37 At one point, one even hired a consultant *on the sly* to evaluate what we were doing!

38 I describe my immediate boss as "such a 'family substitute' on these [work] trips. A nice person whom I value" (journal, 7 November 199X). On another occasion, he said "Did I ever tell you I'm glad you're on the project?" (journal, 5 February 199X).

39 In a recent visit to FRI, one of my longstanding Northern heterosexual men friends, a colleague in a sister institution, initially approached me with caution, not giving me the hug he probably would have in the past. He knew I held feminist concerns, and he acknowledged that the me-too movement had made him skittish.

40 His role as DG required frequent plucking of the political harp string, something the rest of the staff did most consistently in relations with partners in host countries.

41 Other important themes I considered included disciplinary distinctions, particularly those between biophysical and social sciences (far less intrusive at FRI than at other centres under FRI's umbrella organization), top-down vs. bottom-up approaches (all of which were routinely attempted) and FRI's location on the academic vs. applied spectrum. Officially FRI was supposed to produce 'strategic' research, lying mid-way between these latter two poles.

42 I went to Central Africa in the early years:

> You could easily come here and fail to realize the disparity between what the whites have and what the blacks have. The same salary discrepancy here as in Indonesia. Men who really want the jobs discussed (ours, WWF's) are dressed in their best (shabby suits) and their hunger is palpable. . . . People should not have to *want* like that. I wanted the [FRI] job in 199X, but the fact that I didn't *need* it vastly reduces the importance of my own desire.
>
> *(journal, 4 April 199X)*

43 Not that such attitudes were *limited* to Central Africans!

44 Connell (1995) refers to a "pattern of a technical peer group sustaining a strongly masculinized definition of expertise" (p. 181). I have already noted the tendency of professional women in such contexts to take on attitudes and skill sets defined as masculine. The higher the woman's status, the more likely such a shift would have occurred, or a natural inclination been realized—though this necessity may well be changing under the more equitable gender relations of recent years.

45 See Colfer (1983), which discusses the ways in which narratives of inequality effectively reduce a person's perceived articulateness, in Iran, in the forests of the US and in academic discourse, or more recently, Colfer, Sijapati Basnett and Ihalainen (2018).

46 Of course, FRI scientists were also genuinely interested in discovery, in learning new things, in producing knowledge that could be used constructively. Issues of masculinity were only one lens—the one emphasized here—through which to view their behaviour and goals.

47 My time was spent in Brazil, Peru and Bolivia and my knowledge of Spanish was minimal, of Portuguese non-existent.

48 I worked most in Cameroon, with shorter stays in Burkina Faso, Côte d'Ivoire, Gabon, Kenya, Madagascar, South Africa, Tanzania, Uganda and Zimbabwe. French and English were the languages I used.

49 My work in Indonesia predominates, but I also spent time in Laos, Malaysia, Thailand and Vietnam. I speak Indonesian (making Malaysian intelligible) and have varying degrees of fluency in several other Indonesian languages; no Lao, Thai or Vietnamese.

50 See Arora-Jonsson (2009) for the incompatibilities between Swedish and Indian women's vs. men's ways of organizing and relating to forests.

51 See Mabsout and Van Staveren (2010) for evidence of employed Ethiopian women more assiduously performing household tasks to 'make up for' their intrusion into the masculine sphere of wage labour.

52 In Sitiung, West Sumatra, I wrote in my journal,

> [an Indonesian partner soil scientist] came by, and we discussed unfairness. I have tried not to think about that. I truly think it makes me sick [I *was* sick]. But . . . it still comes unbidden . . . Is that why I cry?
>
> *(journal, 10 September, 198X)*

53 Among the environmentally oriented men Connell (1995) studied, "[one man's] renunciation of his masculine career was a highly masculine act. Among other things, he did not tell his wife about it until after he had bought their farm". Similar events occurred at FRI.

54 In the FS-S project, so far from government oversight, we were able to make creative and professional use of the talents available from our spouses, though many of their contributions were unpaid.

55 I was not innocent in this issue: In my own household, where we paid more than most of my colleagues to our household staff, I was paid around 100 times what I paid my household workers. This differentiation was almost as shockingly inequitable as that between low-paid American workers vis-à-vis the now-infamous 1%.

56 A few of us made special efforts to be fairer, paying more, organizing health insurance and paying for schooling, but such behaviour was 'rowing upstream' against the 'development set' context in which we lived.

57 One scientist reviewer, a man, said: "Often I was confused because males have so much dominated outside-the-home work life, that I can't separate the male from the dominant". This is true, as many scholars have pointed out—what is masculine is what is considered 'normal', particularly in these contexts where there are so many more men than women in positions of power and where the subject of concern is a sphere dominated by men and manliness in the global North.

58 Although I do see progress, it is important to acknowledge that I also still heard stories in 2019, showing that sexism remains alive and well in the forestry world. We must further analyze and correct the elements of masculinities (and femininities) that reinforce such experiences.

59 Bribiescas (2016) notes, "As the importance of physical strength declined [evolutionarily], knowledge and experience took on more central roles in the daily lives of older men" (p. 11).

References

Anderson, Eric. 2009. *Inclusive Masculinity*. New York: Routledge.

Arora-Jonsson, Seema. 2009. "Discordant Connections: Discourses on Gender and Grassroots Activism in Two Forest Communities in India and Sweden: Winner of the 2009 Catharine Stimpson Prize." *Signs* 35 (1):213–240.

Boylan, Jennifer Finney. 2008. *I'm Looking Through You: Growing up Haunted: A Memoir*. New York: Broadway Books.

Brandth, Berit, and Marit S. Haugen. 2000. "From Lumberjack to Business Manager: Masculinity in the Norwegian Forestry Press." *Journal of Rural Studies* 16:343–355.

Bribiescas, Richard G. 2016. *How Men Age: What Evolution Reveals about Male Health and Mortality*. Princeton, NJ: Princeton University Press.

Chopra, Radhika. 2011. "Local Lives, Global Dialogues: Shifting Discourses of Masculinity in India." In *Men and Development: Politicizing Masculinities*, edited by Andrea Cornwall, Jerker Edström, and Alan Greig, 139–152. London: Zed Books.

Clarke, Paula K., and W. Ted Hamilton. 2013. "The Value of Anthropology: Noticing and Challenging the Magic Counting Dragon." *Anthropology News*:31.

Colfer, Carol J. Pierce. 1983. "On Communication Among 'Unequals'." *International Journal of Intercultural Communication* 7:263–283.

Colfer, Carol J. Pierce. 2005. *The Complex Forest: Communities, Uncertainty, and Adaptive Collaborative Management*. Washington, DC: Resources for the Future/CIFOR. book manuscript.

Colfer, Carol J. Pierce. 2006. *Vignettes from a Year in Borneo: Local People and Conservation*. Lulu.com.

Colfer, Carol J. Pierce. 2013. "The Ups and Downs of Institutional Learning: Reflections on the Emergence and Conduct of Adaptive Collaborative Management at the Center for International Forestry Research (CIFOR)." In *Adaptive Collaborative Approaches in Natural Resource Governance*, edited by Hemant Ojha, Andy Hall, and Rasheed Sulaiman V, 48–101. London: Routledge.

Colfer, Carol J. Pierce, Bimbika Sijapati Basnett, and Markus Ihalainen. 2018. "Making Sense of 'Intersectionality': A Manual for Lovers of People and Forests." *CIFOR Occasional Paper* 184:40.

Colfer, Carol J. Pierce, Reed L. Wadley, Emily Harwell, and Ravi Prabhu. 2001. "Assessing Intergenerational Access to Resources: Using Criteria and Indicators in West Kalimantan, Indonesia." In *People Managing Forests: The Links Between Human Well-Being and Sustainability*, edited by Carol J. Pierce Colfer and Yvonne Byron, 190–213. Washington, DC: Resources for the Future.

Colfer, Carol J. Pierce, Reed L. Wadley, and Enis Widjanarti. 1996. "Using Indigenous Organizations from West Kalimantan." In *Indigenous Organisations and Development*, edited by D. Michael Warren, Peter Blunt, and David Marsden, 228–238. London: Kegan Paul International.

Colfer, Carol J. Pierce, Joe Woelfel, Reed L. Wadley, and Emily Harwell. 2001. "Assessing People's Perceptions of Forests: Research in West Kalimantan, Indonesia." In *People Managing Forests: The Links Between Human Well-Being and Forests*, edited by Carol J. Pierce Colfer and Yvonne Byron, 135–154. Washington, DC: Resources for the Future.

Connell, R. W. 1995. *Masculinity*. Berkeley, CA: University of California Press.

Dove, Michael R. 1999. "Representations of the 'Other' by Others: The Ethnographic Challenge Posed by Planters' Views of Peasants in Indonesia." In *Transforming the Indonesian Uplands: Marginality, Power and Production*, edited by Tania Murray Li, 203–229. Amsterdam, Netherlands: Harwood Academic Publishers.

Eriksen, C. 2013. *Gender and Wildfire: Landscapes of Uncertainty*. New York: Routledge.

Fine, Cordelia. 2010. *Delusions of Gender: How Our Minds, Society and Neurosexism Create Difference*. New York: W. W. Norton & Co.

Fowler, Cynthia. 2012. *Ignition Stories: Indigenous Fire Ecology in the Indo-Australian Monsoon Zone*. Durham, NC: Carolina Academic Press.

Freeman, Derek. 1979. "Severed Heads that Germinate." In *Fantasy and Symbol*, edited by R. H. Hook, 233–246. London: Academic Press.

Hancock, Graham. 1989. *Lords of Poverty*. London: MacMillan (A Mandarin Paperback).

Harding, Sandra. 2004. "A Socially Relevant Philosophy of Science? Resources from Standpoint Theory's Controversiality." *Hypatia* 19 (1):25–46.

Hearn, Jeff. 2011. "Gender Regimes Changing Men or Men Changing Gender Regimes? Challenges for National and Transnational Social Policy, Gender Equality and Organizing with Men." In *Men and Development: Politicizing Masculinities*, edited by Andrea Cornwall, Jerker Edström, and Alan Greig, 155–169. London: Zed Books.

Hempenstall, Peter. 2017. *Truth's Fool: Derek Freeman and the War Over Cultural Anthropology*. Madison, WI: University of Wisconsin Press.

Hendriks, Thomas. 2014. "Race and Desire in the Porno-tropics: Ethnographic Perspectives from the Post-colony." *Sexualities* 17 (1/2):213–229.

Hendriks, Thomas. 2017. "A Darker Shade of White: Expat Self-making in a Congolese Rainforest Enclave." *Africa* 87 (4):683–701. https://doi.org/10.1017/S0001972017000316.

Katila, Saija, and Susan Meriläinen. 1999. "A Serious Researcher or Just Another Nice Girl?: Doing Gender in a Male-Dominated Scientific Community." *Gender, Work and Organization* 6 (3).

Katila, Saija, and Susan Meräinen. 2002. "Metamorphosis: From 'Nice Girls' to 'Nice Bitches': Resisting Patriarchal Articulations of Professional Identity." *Gender, Work and Organization* 9 (3):336–354.

Killick, Andrew P. 1995. "The Penetrating Intellect: On Being White, Straight and Male in Korea." In *Taboo: Sex, Identity and Erotic Subjectivity in Anthropological Fieldwork*, edited by Don Kulick and Margaret Willson, 76–106. London and New York: Routledge.

Li, Tania. 2007. *The Will to Improve—Governmentality, Development, and the Practice of Politics*. Durham: Durham University Press.

Mabsout, R., and I. Van Staveren. 2010. "Disentangling Bargaining Power from Individual and Household Level to Institutions: Evidence on Women's Position in Ethiopia." *World Development* 38 (5):783–796.

McGinley, Ann C. 2004. "Masculinities at Work." *Oregon Law Review* 83 (2):359–434.

Merrill-Sands, Deborah. 1997. "1997 CGIAR Human Resources Survey: International Staffing at the CGIAR Centers with a Focus on Gender." *CGIAR Gender Program Working Paper* 15. Washington, DC: CGIAR.

Moro, Juan Camilo. 2018. *Borneo Dammed.* Middletown, DE: Juan Camilo Moro.

Pascoe, C. J. 2007. *Dude, You're a Fag: Masculinity and Sexuality in High School.* Berkeley, CA: University of California Press.

Reed, Maureen G. 2003. "Marginality and Gender at Work in Forestry Communities of British Columbia, Canada." *Journal of Rural Studies* 19:373–389.

Reed, L. Wadley, and Carol J. Pierce Colfer. 2004. "Sacred Forest, Hunting, and Conservation in West Kalimantan, Indonesia." *Human Ecology* 32 (3):313–338.

Reed, L. Wadley, Carol J. Pierce Colfer, Rona Dennis, and Julia Aglionby. 2010a. "The 'Social Life' of Conservation: Lessons from Danau Sentarum." *Ecology and Society* 15 (4).

Reed, L. Wadley, Carol J. Pierce Colfer, Rona Dennis, and Julia Aglionby. 2010b. "'Paper Park' and the Social Life of Conservation: Lessons from Danau Sentarum." *Borneo Research Bulletin (Special Issue)* 41.

Saini, Angela. 2017. *Inferior: How Science Got Women Wrong—And the New Research That's Rewriting the Story.* Boston, MA: Beacon Press.

Yasmi, Yurdi, and Carol J. Pierce Colfer. 2010. "Forestry and Fishery Conflict in Danau Sentarum: Application of an Impairment Approach." *Borneo Research Bulletin, Special Issue on Danau Sentarum* 41:145–161.

Yasmi, Yurdi, Carol J. Pierce Colfer, Linda Yuliani, Yayan Indriatmoko, and Valentinus Heri. 2007. "Conflict Management Approaches Under Unclear Boundaries of the Commons: Experiences from Danau Sentarum National Park, Indonesia." *International Forestry Review* 9 (2):597–609.

7

REVISITS TO OLD HAUNTS

With Andrew Balan Pierce and Sufiet Erlita[1]

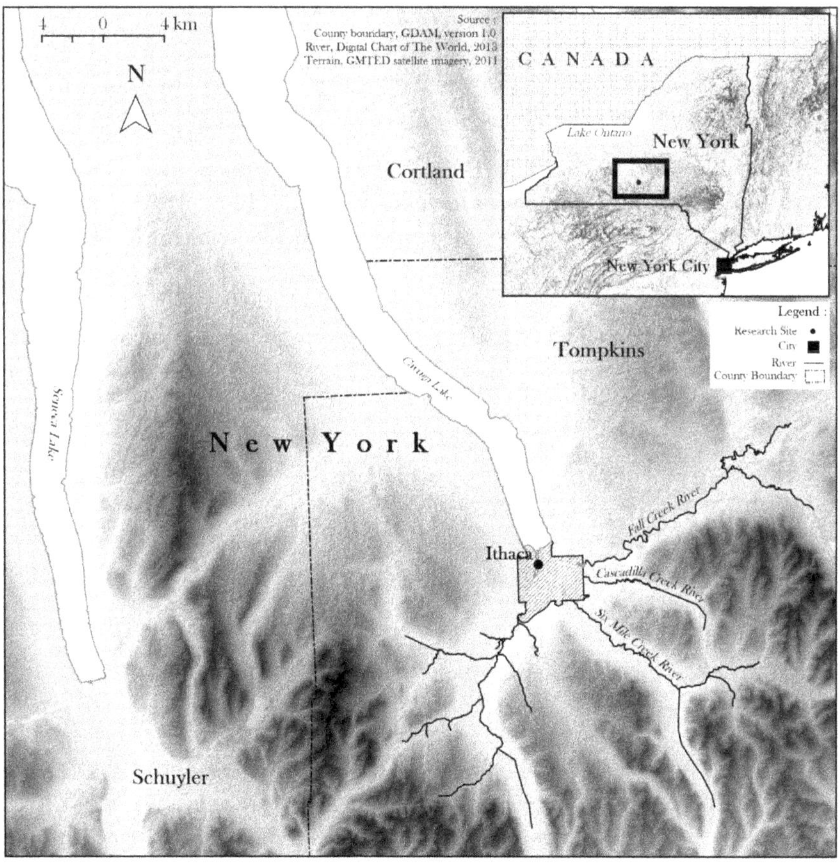

FIGURE 7.0: Map of Ithaca

Introduction

This chapter focuses on masculinities in the recent and current world. I begin with a return to the reflexive element begun in Chapter 2, because of the particular difficulty—as discussed previously—of providing comparatively objective assessments about gender when one is inevitably enmeshed in one's own gender system. Thinking we can differentiate genuine transformations in a particular community from more global changes that *affect* that community or transformations in the researcher's perspective due to age and experience is a chimera. The best we can do is be clear about our own perspective and experience.

I then turn to what I observed in May to June 2017 in Bushler Bay, the one-time American logging community (Chapter 3). That discussion is followed by similar observations in the two Indonesian villages of Long Segar and Long Anai (East Kalimantan), in March 2019 (Chapter 4). I make a few observations about conditions in Sitiung, West Sumatra, based on work organized by Dr. Ardi from Andalas University in Padang. These are followed by brief concluding remarks, expanded in Chapter 8.

Reflexive ruminations on American and global masculinities in 2019

Gender systems have changed markedly since the 1970s. At the global level, there have been regular international conferences, policies and action plans since 1975, raising gender's visibility and supporting more equitable relations between men and women.[2]

In the US, though the current political climate includes considerable 'backward motion', there have been great strides forward in terms of gender equality over my lifetime:

- Women's reproductive rights were reinforced by legal precedents (e.g., *Roe vs. Wade*, 1973).
- Birth control became widely available (e.g., via Planned Parenthood, commercial markets, sex education in schools and better medical advice to patients).
- Women's sports opportunities were greatly expanded through Title IX[3] (enacted in 1972).
- Women's involvement in formal employment increased and became more socially acceptable.

In all these changes, however, the emphasis was on women as 'the disadvantaged'. These changes also have had implications for men, manhood and masculinities. Thinking in terms of conventional men's roles, smaller families have meant lower levels of required responsibility for men as providers, which can be seen in a positive light or, more negatively, as a reduction in men's *importance* (seen in some examples from global GENNOVATE studies; Elias et al. 2018). Women in sports requires a rethinking of the meaning of sports as a symbol of masculinity among many groups, as a mechanism by which physically talented men gain prominence over others (e.g., Anderson 2009 or Connell 1995). Similarly, women in the labour force have required men to adapt, to face challenges (whether explicit or implicit) to their earlier roles, often as sole provider for the family (e.g., Reed 2003, on men and women in Canadian logging communities; or Petesch et al. 2018, for sub-Saharan Africa).

Here, I shift from these broader societal changes to the changes I've experienced in my own relations with men, looking at specific relationships and the contexts in which they have played out. Besides providing a glimpse into my own biases and assumptions, these patterns are not atypical of middle-class, American, highly educated, white couples in university neighbourhoods (in this case, Cornell University). Clearly *not* representative of American life at large, these patterns nicely represent the population likely to be active in international circles, including those that pertain to forest-related policymaking (see Nader 1972, for a dated but still timely perspective on the need to 'study up').

My husband, Dudley, has provided my most intimate current knowledge of American masculinity. We married in 1985 (in Sitiung); he is of middle-class background and has a PhD in fisheries biology. Like my first husband (see Chapter 2), Dudley is from the American Northeast and has been supportive of my professional aspirations and passions. I have also worked collaboratively with him on professional projects.

He fulfills a number of conventional ideals of American masculinity. He is (or was—like me, he shrank a bit) six feet tall, attractive, was good at sports before knee problems prohibited that activity, knows a lot about fixing things around the house (indeed, likes to) and is the 'strong, silent type', tending to keep his feelings to himself (sometimes unaware of them). He was the main provider in his previous marriage. He was an early student of computer languages (substituting one of those for the language requirement in grad school!), has always been good at math and science and values logic and knowledge highly.

On the other hand, he knows how to cook (though I cook most meals) and occasionally enjoys it, willingly performs various household tasks and has fairly cheerfully been a 'trailing spouse' for most of our married life. When I offered periodically to leave the job I loved, he consistently declined, expressing his appreciation of the intense work involved in periodic consulting and the full-time breaks between assignments. He also recognized the value of my comparatively secure employment and sufficient income. In his retirement, he is a gardener, a sailor, a photographer and an eternally curious individual. Like me, he is peripherally associated with Cornell University (as an 'adjunct associate professor'—a somewhat more prestigious title than mine, 'visiting scholar').

Although I believe he feels comfortable in his masculinity, I don't believe that it is of central concern to him (though perhaps it was when he was younger?). I have not seen him trying assertively to demonstrate it, as some men do; nor have I seen him responding defensively to what might be perceived as threats to it; he is more likely to give a dismissive snort. The fact that he has not been the primary provider in our marriage occasioned periodic disapproval from his mother in the early years. Once in the 1980s when I mentioned to her my view that when a husband and wife were both employed the domestic tasks should be split, she responded with shock and dismay, 'You don't expect him to do *housework*, do you!?'. She would also sometimes express her feeling that he should 'get a real job'.

Although never to my knowledge subjected to the snide harassment my former husband experienced in Bushler Bay, Dudley was irritated sometimes by my FRI co-workers, a few of whom seemed to dismiss his inputs because he was a spouse rather than an employee (even when he was occasionally working as a consultant there). But I saw no evidence that he took it to heart as some men would. The degree to which this is a personal trait or a change in both the general atmosphere and a personal adaptation to that remains a question.

He is interested in and enjoys sexuality, but if it is a defining feature of his masculinity (as it was for my previous husband), that is not clear to me. Nor is his enjoyment of sports. He watches football every fall, enjoying particularly the links with the University of Georgia, where he worked for eight years before I knew him. But when we lived overseas and could not watch it, he did not seem to suffer (unlike my father);[4] nor does he watch any other sports except soccer World Cups (both men and women). He is not violent, and has never beat me, my children or anyone else, to the best of my knowledge.

We have a 'companionate marriage', with what has evolved to be ironically a somewhat traditional division of labour:[5] I now cook, shop, wash dishes and clothes, occasionally dust surfaces and clean two of the bathrooms. He does the floors, windows, supplementary cleaning in the kitchen and bathrooms and takes complete charge of our three acres of land—the vegetable garden and most of the flower garden, organizing mowing and some snow clearance, reducing invasive species and doing virtually all outdoor tasks. He also takes care of his own needs during the three months of the year that I am away caring for my aged mother.[6] Why the somewhat traditional division? Most simply, because we have gendered expertise gained during our youths, combined with his greater physical strength, required for some of these tasks. The other couples our age whom I know well show similar tendencies: attempts to ensure gender equity, but in some cases lack of the knowledge, experience and physical strength (and perhaps the motivation) necessary to totally shift traditional tasks. As upper-middle-class women, many hire someone to do the heavy house cleaning tasks (floors, bathroom cleaning, windows, yard work) typically organized by the woman.[7]

The masculine harp strings that characterize the older men I know well include breadwinner, father/grandfather, fixer of household equipment, manager of vehicles and heavy equipment, caretaker and mathematics (in the form of taxes, bill paying, etc.). But I see a much wider spectrum of publicly acceptable and visible gender arrangements now vis-à-vis most of the 20th century.

BOX 7.1 SEXUALITY AMONG THE AGED—RUMINATIONS FROM ITHACA

I was driving home on New Year's Eve, from dropping off my husband's walker and the ice machine we'd borrowed to use during his knee replacement (which had morphed into a heart attack) from the Finger Lakes Independent Living Center a month before. Stopping at a red light, I found myself next to a Kenworth truck. Looking up from my tiny Prius at the huge vehicle beside me, I was transported briefly back to my time in Bushler Bay, flooded with the memory of my own passion (now so dimmed). The big truck loomed over me, symbolic of a certain kind of in-your-face masculinity, which I'd found embarrassingly attractive in my youth, and still somehow remained attracted to, despite my intellectual recognition of its toxic elements. Such symbolism works through the emotions, not the intellect.

Changing gender roles and ideals appear even more obvious among younger, educated couples. Some marry, some don't. Some are heterosexual, some are not. Some follow traditional gender roles in their households; some switch roles; others create unique patterns based on personal and shared preferences. Occasionally husbands stay home and take care of children while women go out and work. One such employed wife mentioned her theory that some women who stay at home with their children look at childcare as a kind of profession, striving for 'success' in much the same way that formal workers may. When I recounted this theory—which I also found believable—to a stay-at-home dad, he agreed that he felt that way about fatherhood.

The fathers who stayed home though felt some dis-ease about not being the breadwinner; two expressed the view that if they were to continue emphasizing fatherhood in this way, they would likely be subject to more pressure from extended family members to return to breadwinning. In general, all those I know well—men and women—strive to be egalitarian in their division of labour.

I do not mean to suggest that American marriage has become entirely or uniformly equal. I live (we all live) in clusters of people who share many traits (such as the relatively elite 'island' that is Cornell University), and I know there are strong trends about in America (some in the rural world encompassing my university), trends that encourage stricter gender dimorphism, celebrate traditional gender roles and decry egalitarianism and inclusivity—symbolized by current President Trump.

Turning to my four children, the oldest (born in 1969) did not complete college and is in a somewhat conventional marriage, with her husband as wage earner and she a homemaker (a role she sought). She has taken on motherhood as a profession and both she and her husband have strong voices in household decision-making. Her husband plucks the breadwinning, fatherhood and sports harp strings most consistently—spending his free time coaching the mainly boys' hockey team, for which one of his daughters is goalie; he rarely does any housework, beyond cooking an occasional meal.

Our other three children approach domestic tasks more equitably than would have been common in my youth. One son (PhD in history) works as an international intelligence analyst, keeping him away from home 10–11 hours/day (>two hours/day commuting). When confronted with a choice, his job or hers (same degrees) in different states, they chose his. But he is actively involved in most household tasks, including childcare, insofar as his job allows—far more involved than would fathers have been in my youth. Fatherhood, caring and breadwinning are his harp strings most visible to me (though at work I believe he replicates many of those described in Chapter 6).

My other son has a BA and his wife, some college; they have no children yet. Both are of an entrepreneurial bent that combines business with the arts; they also share household and work tasks quite evenly—though they have recently lived in Indonesia, where household labour can be obtained inexpensively.[8] I recall my now divorced daughter's description of their household division of labour: "I do the wet things, he does the dry". None of our children fully accepts the traditional American ideas about how marital relations and labour should be conducted/structured that I grew up with (and rejected).

With regard to older couples in the US, I offer what is basically a series of hypotheses based on my experience in rural Washington and New York[9] states, and among my own relatives all over the country. Among the older, middle-class, American couples I know well, women seem to have gained authority, though also probably work harder vis-à-vis their husbands.[10] The women are usually somewhat younger and tend to be in better health. Husbands, when they retire, enter a domain that has typically still been their wives' to manage, the household—a process that requires adaptation (e.g., Hatch 2000; or Varley and Blasco 2001, on urban Mexican men). At the same time, the broader context (mass media, education, policy)—both American and international—has been marked by increasingly serious attacks on male dominance and patriarchy as ideologies and institutions. Women, who have accepted some version of male dominance throughout their lives, are unlikely suddenly to change their long-established interaction patterns; but neither are they likely to accept their husbands' right to determine life as much as they may have in the past. The outcome seems to be—and again this is more a hypothesis than a 'finding'—a small take-over by the women, who now may drive, organize social events and express more independence than previously. When I've expressed this view, a couple of women have argued that it's always been like that, that it's not a change occurring in old age. Maybe.

Observations from Bushler Bay in 2017[11]

In 2016, I began planning a brief re-study in Bushler Bay (Figure 3.1). I had kept in good, mostly long-distance contact with the man I'd been sexually involved with in the 1970s (his wife had died a decade earlier and he had remarried). I became 'friends' on Facebook with several other friends from the 1970s. One coterie of these folks had been John Birch Society sympathizers (a right-wing political group) and are now avid Trump supporters. I've regularly exchanged conflicting political views with them, in my effort to understand their perspective (though I'm not sure it's helped much)—mostly after the fieldwork reported in this section. Many Bushler Bay residents, however, lean to the more 'progressive' end of the American political spectrum.[12]

Given the changes I'd seen in upstate New York, and my many years away from the US, I was excited to see how this once-familiar setting had changed. I made a quick, two-day trip in the spring of 2016, finding that superficially, it looked much the same: a tiny cluster of businesses around the school as the centre of town, a main highway running through it. I informally interviewed a few people, which whetted my appetite for more.

The study Hummel (an ecologist) and Cerveny (an anthropologist), both USFS employees, and I planned focused on women's involvement in forest management. They also wanted to test the potential of rapid rural appraisal (RRA) tools for use by the US Department of Agriculture's Forest Service (USFS) research branch (also the funding agency). I spent two weeks in Bushler Bay in May to June 2017 and tested six RRA tools as well as conducting what I call 'ethnography light' (an approach built on close and longstanding links with the communities). The tools tested included two pebble sorting tools (one on non-timber forest products and one on forest activities), the Who Counts Matrix, an interview guide, participatory mapping and future imaging, plus a conventional questionnaire—all administered using pre-existing relationships, emerging opportunity and snowball sampling.

Several studies had been done in the area, in addition to my own. Informal harvest of forest products was and remains an important activity. Both communities were directly affected by the 1990s 'timber wars' (Bari 1994; Dark 1997). Since the near-demise of logging in the area, Olympic Peninsula counties have experienced higher unemployment and population decline along with changing identity (Kirschner 2010). Of the in-migrants, many are 'amenity migrants'.[13] The beauty of the area, its rainforests and beaches, remain a steady draw for visitors. The forests, which once supported the communities financially, are increasingly appreciated for scenery, heritage, outdoor recreation and tourism.

Driving into Bushler Bay in 2017, I was again struck by how little had changed. Besides the school, the small businesses and the highway, the USFS office remained on the way south out of town. Some of the residential areas near town were a little more dilapidated, but further out, there were some homes, much more luxurious than in earlier times. The population was a little lower these days (from 500 or 600 in the 1970s to 437 in 2015). As I dove into the research, I discovered that life was really quite different now.

One of my earliest experiences this time around was a run-in at the café, where loggers always used to assemble looking for the day's work:

> When I walked into Bushler Bay's cafe, Hubert[14] was there (I don't know him, but he was dressed as a logger). I announced happily that I was coming back to Bushler Bay after 40 years and would be here for two weeks. He wanted to know my name, and I told him. He said "Oh, you're *that* woman. I hear you wrote some pretty bad things about Bushler Bay". I expressed surprise and asked if I could join him, and he said, "No", point-blank. Then I asked his name, and he said, "That's for me to know and you to find out". The waitress harassed him about being so rude, called him an asshole even. But it was clear they are friends. I explained that things were pretty bitter between the loggers and the Forest Service in those days and that we'd tried to reflect the feelings each group had for the other. He began to warm up a bit. He said, "The Forest Service folks, they always felt they were better than us, and that don't fly around here. Some of them were alright, but the higher-ups, they were the worst". He was also under the impression that the "Forest Service don't even replant anymore" [untrue]. Turns out he came here in the 1970s from Illinois, started out as a choker-setter and ran a lot of heavy equipment. He'd done almost all the logging jobs, but had multiple surgeries, at least some on his back. When I asked if he was retired, he said, "The state retired me. They got tired of paying for my operations". He also said, "Dan [a respected Local[15] long-time logger] says 'they're killing our kids'". (Not sure if 'they' refers to the USFS or the government or what). So I'm suspecting Dan may be the source of the story that we wrote bad things. I'm pretty sure Dan read the case study.[16] So. . . . getting words out of one-time loggers aint gonna be easy!
>
> *(notes, 23 May 2017)*

> Just got back from a trip to town, for a meal at the café. Learned that the Pecker[17] Pole micro-brewery is in the back garage of the café. T-shirts with the Pecker Pole logo available for $20.
>
> *(notes, 22 May 2017)*

The interchange with Hubert reminded me of the challenges of doing fieldwork in this community, where privacy was so central. The second observation suggested that publicly, sexuality remained an appropriate topic, focused specifically on men's bodies (as in the 1970s). Hubert's comments represented a nice intro to several recurrent masculine harp strings—disdain for the 'chords' represented by the USFS (social hierarchy, concern about job security, indoor work), a strong value on independence and straight speaking, commitment to the danger and skills required in logging and its role as a key chord in Local Bushler Bay 'songs'.

The environmental pressures that had begun in the 1970s had continued, resulting in a halt to all public logging near the community. Some private logging continued.[18] But logging was no longer a primary source of the community's livelihood; only five loggers were said to remain. The conflict between loggers and the USFS had also dissipated. There were few USFS personnel still working there and the Washington State Shellfish Lab had closed down, both reducing the number of Public Employees. The source of internal conflict now, though reduced, was between the environmentalists and those earlier labelled Locals. In one interview with a community leader (an in-migrating man), he

> began by saying that the community was divided in two between the 'older pioneers, lumberjacks' who remember a time when there was lots of work, three gas stations. Enter the spotted owl.[19] One of his neighbours' father went to the mill,[20] was laid off, and came home and blew his brains out.
>
> The other half is what are termed 'environmentalists' or 'tree huggers'. The two groups have nothing to do with each other. Won't work together. But the lumberjacks are dying off and environmentalists are increasing, so they will 'win', he said.
>
> *(notes, 25 May 2017)*

Another key change was a simple demographic one. Whereas in the 1970s, the age distribution had been a fairly normal pyramid (with 12% of the county over 60), now between 31% (Bushler Bay proper) and 48% (the nearby Hood View[21] hamlet) were in that age range. Many of these elders are in-migrants from urban areas across Puget Sound and elsewhere. These facts have meant a huge reduction in the significance of the breadwinner harp string: few jobs available and many people past working age.

In the survey we conducted at the Hood View shrimp festival (n=21, diverse ages), we asked "What kinds of things do you do in the woods for income?". Of those who answered, 38% of the women and 63% of the men reported no forest-related income, whereas 24% of women reported some (related to tourism, an environmental NGO and as a stimulus to creativity). The only men reporting any such income were 13% who reported having *previously* planted trees. A quarter of men and women left the question blank, which may have meant they also gained no income from the forest.

Looking at gender, the most dramatic change was the reduction in its apparent significance as a differentiating social structural feature. The clubs in the community, previously strictly gender-segregated, are now integrated. The Who Counts Matrix[22] strove to distinguish the importance of men and women for forest management on seven dimensions. The dimensions[23] were assessed on a scale of 1 (high relevance for forest management) to 3 (low relevance). On average (means), for Retirees (2.2) and Locals (1.3), men and women's scores were the same, and for Public Employee women and men there was only a 0.2 difference (2.0 to 1.8, respectively), with men considered to 'count' slightly more. The biggest difference was for 'pre-existing rights', which were considered greater (lower number) for men than for women, across the board.

The pebble distribution methods confirmed both the reduction in gender differentiation and the diminution of importance for logging. Thirteen activities (previously determined to be of relevance in the region) were assessed by 23, mostly elderly people,[24] distributing 100 pebbles among the activities. Had we used such a tool in previous times, logging would surely have gotten a very high proportion of the pebbles, leaving only a few for the other activities. However, the 2017 results were fairly evenly divided among activities, with six activities getting exactly the same assessment for men and women. With the exception of logging, for which women were allotted 1.5 pebbles to men's 4, and hunting (2.5 for women, 5 for men), no other activities were differentiated gender-wise by more than 1 pebble (or 1%). The harp strings associating men with logging and with outdoor activity remain, but are seriously eroded, especially if seen as contrasting with women's involvement (a one-time reinforcement of masculine dominance).

The pebble distribution tool on non-timber forest products, which assessed men's and women's involvement in only seven categories,[25] showed more gender differentiation, but men and women were both involved in collection of all the products. The average allotment of pebbles was 7.1 per category, with men averaging 8.3 and women, 5.9. As with activities, these differentiations are not extreme, suggesting a lack of significance in terms of masculine harp strings—or perhaps a trend toward the muting of gender difference. A question remains in my own mind as to whether these findings derive from the elderly sample (some have argued that gender differentiation reduces in old age)[26] or from the broader societal reduction in emphasis on gender differentiation.

To examine the current manifestation of the harp strings that were important in earlier years, we turn to participant observation and the rich interview data (interview dates, given below, once for each man, apply to all associated quotes). Here I recount the varied stories of three men: one a Local man (whom I'll call Tom), who has adapted to the economic changes and who retains a quite traditional view of gender roles, with 'modern' variations; one a working in-migrant (Dick), with a lifelong history of visiting the area before moving there; and another native son (Harry), who is now seriously handicapped and semi-retired.

Tom, of middle age, comes from a very religious natal family in which his father ruled the roost.[27] The family was poor and the father worked long, difficult hours logging to make ends meet. Tom and his brother, on the other hand, feel able to make more money with fewer hours and easier work.

> Each owns two homes, gets up late and is better off than their dad. Both of their "wives are happy with a credit card, plenty of gas, a working car". His wife could go help [a relative] in another town, she could afford to do it.[28]
>
> *(notes, 29 May 2017)*

There was a clear element of competition between sons and their father.

Tom expresses his belief in and practice of traditional gender roles. He says, "I think of myself as the provider-protector of my family"—two conventional masculine harp strings expressly emphasized several times. He sees men and women as having different, but complementary natures: The man manages (and presumably controls): "I handle all the money, administer the household, have the keys to the post office". But in the same breath, he acknowledges that this is a "lost lifestyle in the modern world". He also emphasizes that his role and his wife's are of equal importance.

> "Women are inherently better caretakers, sometimes to a fault. I live in a traditional family. I say we have two incomes; I have both the jobs. My wife is a wonderful wonderful wonderful person. She is raising the kids."
>
> He feels he's the protector/provider for his wife. But, he says, it's not a lopsided relationship. He respects his wife. On the balance sheet, he provides the money, and she provides a wonderful home. Their work is of equal value.

He reminds me that he's not a logger, but that he's able to use his logging skills (from previous days) in his successful company. He has "felled trees professionally, driven a log truck, loading, lots of logging jobs in earlier times". He says, "Men want to be respected, to feel strong; the logger and outdoorsman are still important ideals here. Even when we're old, men want to be strong".

His interest in fatherhood and, even stronger interest in being a protector, come out in several statements. Where his wife likes "the idea of a wolf howling in the distance", he would prefer to get rid of wolves, focusing on the dangers they represent. He remembers one tearing a beloved dog to shreds. He feels his wife doesn't realize the danger they pose.

His emphasis on rationality is evident from several of his comments. He interprets his maintaining an orchard where deer come to browse as "chumming for cougars and coyotes". He says he hunts deer partly to reduce this population that draws bears and cougars into his terrain. Elsewhere, he emphasizes the importance of balance, between men and women, in extraction vs. protection of forests, in the economy:

> In the '50s and '60s, even the 1970s, a logging company would have two timber fallers on one tree, two skidders at the landing, two equipment operators—8–12 people grouped on a landing. Six loads of trees might go out. Now it's so mechanized that you only need three guys in the forest doing 25–30 loads a day. So, it only supports three families (with the wife working too). Balance is way off.

Tom told a story of one Local family who owned "thousands of acres of forest land". When the owner died without a will, one of his sons assumed that land would go to him as "the only remaining logger, the only one who knew what to do with the land. Loggers don't recognize any other uses". Tom feels that forest lands belong to all the people. "An Indian in Oklahoma also owns this National Forest".

Another significant contrasting aspect of Tom's masculinity, vis-à-vis earlier and more antagonistic Bushler Bay versions, is his attitude about homosexuality:

> In the 1970s there were no social media, less info, just evening news. Every man was struggling to be manly. There was no recognition or tolerance for any other way of being. Now, we know that there are 1000s of gay boys out there, so now there is acceptance. If I was gay, I'd have a boyfriend or husband. When I was a kid, if someone even wanted to be a *chef*, he was shunned or would have been.

Tom also captures Local symbolism reflecting the differentiation between environmentalists (which are lumped with tourists) and loggers—in their preferred kinds of vehicles. Ford F150s, seen as masculine,[29] are preferred by loggers, whereas tourists and environmentalists are seen to prefer Subarus. Subarus are less symbolic of gender, more a set of values and lifestyle, except as distinct from the clearly masculine Ford F150s.

> [Tom] went to the Army . . . and came back in 1995. No stick of Federal Forest was being cut. Only one [forest] thinning sale on Rocky Brook. People [loggers] would build roads to get the trees they cut and they'd maintain them ("for tourists, granola crunchers [environmentalists] and fire suppression"). "Look at the cars: Subarus outnumber F150s 20–1 in the woods". He wants me to go up to Scar Pass and count, or count the cars headed for Hood Canal bridge to see both the kinds and the number. Sometimes they are backed up to Center Road [ca 18 miles], he said. And this traffic has nothing to do with rock quarrying or logging.

Dick, our second example, is a middle-aged in-migrant from one of Puget Sound's cities, with lifelong links to his property, some distance from Bushler Bay proper (notes, 24 May 2017).[30]

> I drove for 25 minutes to get to his house, which is beautifully located on [the water], on the [. . . .] Peninsula. I drove through lovely forests, lining the road part of the way, saw entrancing views of the Olympic Mountains from time to time, as I went by a clear cut; I found him down by the water, [in a beautiful home] down a gravel path, surrounded by forest.

His first words emphasized the changes he'd seen in the local economy: from one based on logging to one based on recreation, including many 'amenity migrants'.

> Logging used to be active. Logging roads were for timber. Now the area is managed for recreation. The Forest Service used to employ a lot of people, now not so. He remembers when he first saw a clearcut and could see the Olympic range. Now they are all up and down the peninsula on

> DNR [Washington State Department of Natural Resources] land. But he watched it grow back. So clearcuts don't really bother him. It will grow back, he's seen it. He's just glad it's not a bunch of houses.

Like Tom, Dick has close ties with his extended family, who share ownership of 40 acres in the immediate area, which had belonged to his grandparents. His love of the forest comes out clearly. Compared to Tom, Dick's sense of his own masculinity is muted. He stressed the importance of management of the forest, of his reliance on government agencies to ensure its sustainability—which implied no antagonism to hierarchy, despite running his own business. Although he recognized the need and legitimacy of logging, he also expressed suspicion about private management, having seen others buy land and promptly denude it, in one case that particularly disgusted him, to buy a motor home.

> "People need to remember how long it takes, he said. It would take over 100 years to grow back. It's a 50–60-year cycle for the DNR. The peninsula was logged around 130 years ago". He saw evidence of the logging, including pegs (or was it notches?) in the stumps (used for the men to stand on when cutting with a cross cut saw) and a logging camp a mile or so up the peninsula from his place.

Dick's relationship with his wife was also markedly different from Tom's. His responses to my questioning what sorts of things he did in the forest follow:

> He hikes, lives in it, gardens, recreates, gets his wood (heating source, which he cuts and his wife splits), drives through it to work, runs on logging roads (to avoid cars), rides his mountain bike, walks the dog, picks mushrooms for food.
>
> He said that his wife does the same things he does in the forest, but she's more sensitive about the remaining trees. He said, "We do everything together. She's my companion and I wanted to introduce her to the things I enjoy". They also ski and climb in the forest—though now he says, he just works. He has [several] employees and business has been booming for the last ten years.

When explaining the maintenance of their widely admired and beautiful garden, their shared experience of caring for it was clear. His parting comment as I left, "I tried to tap my feminine side", reminded me of the cases described by Connell (1995) of young men involved in conservation activism, "where gender hierarchy has lost all legitimacy" (p. 90). I could easily imagine Dick, for instance, saying this quote from one of Connell's cases: "That experience of being alone, wandering round and doing things and appreciating things and enjoying a beautiful place can really give me a wonderfully clear, pure feeling" (p. 128).

His solutions for the problems related to the wildlife he loves are non-violent:

> For himself, he loves the animal life (bear, coyotes, cougars, bobcats, possums, raccoons). The bears have caused him problems, eating his apples (he has an orchard) and honey (he used to keep bees). The solution he's concluded must be a fence, as he doesn't go for hunting. He feels the combination of sea and forest is great for wildlife. A cougar, a few years ago, attacked [his relative's domestic animals]. The cougar had to be shot and she gave up on raising them. He also said coyotes are a problem but didn't elaborate.

Dick's forest-based activities are relevant as Local harp strings for masculinity: outdoors, requiring and maintaining physical strength (e.g., cutting firewood, running), recognizing the value of logging. His company does construction, a conventionally masculine occupation, he lives in the forest (again coded 'masculine') and he maintains and uses heavy equipment, another interest that serves as evidence of masculinity in this region. His activities and skills surely grant him respect from Locals, but his own value system, as evidenced in this discussion, stresses conservation and protection. He'd like to see the nearby Trident Navy base turned into a nature preserve (in rejection of another stereotypical manly institution).

The last example is Harry, a retired man in his 80s who grew up in the community and has lived there all his life, a valued member of the Local community (notes, 23 May 2017).[31] Their house, on the edge of Bushler Bay proper, is small but neat, with a garden full of flowers. Harry has worked as a logger, a USFS employee, in the brush-picking industry and for the US Postal Service. His personal commitment has been to place over occupation. And his comments reflect the usual Local masculine distaste for 'paperwork':

> In the USFS, he found that the higher up he went, the more paperwork. He particularly remembered with distaste the reports he had to file about environmental assessments. He'd write them, and someone higher up might or might not like them. He wound up spending time on paperwork he didn't like. He had a chance to be moved [by the USFS] to California, but his kids were in high school and really didn't want to move, so he didn't go. You had to work in at least two forests at that time.

His children's preference for staying in Bushler Bay (they remain nearby) suggests that fatherhood was important to him. Although he is now unable to walk or talk very well, his love of the outdoors is clear. Indeed, his sense of loss, both from his own disability and from the broader changes that have occurred in the area was palpable, though he expressed none of the bitterness I had anticipated (and heard from some). He listed some of the changes, ones with implications for the Local view of masculinity (see also Colfer 2018):

- Logging was key in the 1970s. In the pebble distribution tool, Harry gave 4 beans out of 100 (all for men) for logging.

- Brush picking was also a going concern for local folks then. Harry now gives it four beans for women and eight for men. This business has been taken over, he says, by Mexicans, Vietnamese and people from Honduras.[32] He understands they have a hard life, expressed sympathy but also aggravation and disappointment.
- Firewood used to be a bigger deal. More people used wood for heat and sometimes cooking in the past. Firewood permits were given out by the USFS after a logging job or a blowdown. There are almost no logging jobs locally from the USFS these days.
- Shellfish used to be readily available to anyone, but now the human population has increased, and people are unwilling to share with others—the land is privately owned (as before), but the norm to let people gather no longer applies.
- Mushrooms are also collected on a broader scale by the Vietnamese (though some do remain for local folks).[33]
- The regulations for fishing are one a day of at least a foot in length; the rule used to be 20 a day, as small as 6" in length.

Although both men and women engaged in all these activities, to varying degrees, they were central to men's outdoors harp string in a way that wasn't relevant for womanhood. In the past, these activities had been important for provisioning and some subsistence, as well.

Harry's love of nature also came out loud and clear when I asked about the places he liked to go. Because of his infirmity we discussed where he'd gone in the past. Looking at a map,

> he picked the Mt. Townsend area and then Buckhorn Wilderness as places he used to fish and hike. He loved those areas and also went there when he worked for the USFS, plus to Gray Wolf River (before these areas were termed Wilderness Areas—now you can only fish and hike). Then he remembered he liked rivers, so he pointed out the Big BB. Then he remembered the Dosewallips which he also likes—both for fishing. [His wife] is from up the Dosewallips, and he remembered that he used to go there to see her before they were married. Obviously fond memories. He said the Dosewallips is a really beautiful river. There was a landslide there that took out 7–8 miles of the road to the [Olympic National] Park. They've been trying to get the USFS to fix the road but so far no luck. I felt I was making him think sad thoughts (sad that he couldn't do these things anymore), so I changed the subject.

He also expressed the same love of wildlife that Dick had shared.

> While we were talking Harry expressed his distaste for people hunting bears. He was ok with [their hunting] deer and elk and used to do that himself for food; but he liked bears, saw them a lot when he was brush

> picking and they never bothered him. He said, "I got to a point where I didn't want to kill things anymore". What a nice, gentle man he is!

Given Harry's age, hunting, fishing and gathering had a serious provisioning function earlier in his life. And he has managed to remain employed for the better part of his life, including some work with the post office even now, suggesting a serious commitment to the breadwinner harp string.

But his wife's active involvement in paid labour now suggests an open attitude—different from Local attitudes in the 1970s[34]—toward a woman's employment, as well as probable financial need. Although I did not see them together on this visit, our interchanges about when to meet reflect an active life of community involvement as well as what must be a significant caretaking role on the part of his wife. His ways of interacting with me and talking about her suggest he does not see his role as 'controller' of his wife. He fits in with her busier schedule.[35]

Other interviews and observations portrayed additional harp strings. Men discussed their children. One widely admired Local described his regret:

> He has some things he regrets, like not spending more time helping his wife with the kids. He's proud of the two youngest, including [name], whom I remember from high school.
>
> *(notes, 2 June 2017)*

The most consistent concern in the community, expressed by all ages and genders, is the lack of employment opportunities for their adult children and those of their neighbours. The men are as likely to express this concern as the women. One girl talked of always going hunting with her father; the pebble distribution results also reflect women's involvement in hunting. Another man reported the bonding function of taking his adult son camping in the Olympic Mountains after a traumatic experience.

The caring/loving harp string pops up again and again. Tom expressed his caring most consistently in connection with his role as protector. Dick talked of his love for wildlife; his care about the forest was obvious. Harry's care about rivers and wild areas was evident, and all three obviously loved/cared about their wives. Many folks mentioned their love for the forest and the environment, making efforts to care for it; some loved their jobs.

Interestingly, especially given its importance in the Bushler Bay of the 1970s, no mention was made of sexuality in all these interviews and informal conversations—either by men or women. I would be surprised if sexuality does not continue to be important to men (e.g., the Pecker Pole beer brand); but there have been news reports about the reduction in sexual activity among America's youth as well as evidence that sex declines in importance as people age—a topic for further research.

A couple of older men mentioned local political action, which in the 1970s had been largely in women's hands, via club activity (Colfer 1977). Now, men's discussions centred on their dissatisfaction with the lack of responsiveness of county government, rather than the very local school levies, provision of street-lights and sports-related issues of old in Bushler Bay proper. There was also an assumption in people's visions of the future that the formal government (USFS, US Park Service, DNR) would and should be responsible for maintaining the forest, for which all expressed a desire. There were complaints about the specific policies and actions of the USFS, for instance, insufficient cutting, insufficient re-planting, too little money, too few workers and too few local jobs. But I heard no aspersions against the masculinity of government workers (many of whom are women now).

Bushler Bay summary

Although there have been very significant changes in Bushler Bay, many of the chords of masculinity remain. Men in their productive years are still concerned to fulfill a provider role; many, including both Locals and amenity migrants, link outdoor work, physical strength, courage[36] and independence in an important chord. Fatherhood and caring remain important for many as well.

But some significant changes have occurred. The most obvious is the change from Local men's sense that they should control their households to a more companionate view of marriage—more closely approaching the previous Public Employee tendency. Yet all the marriages I observed and discussed appeared to value gender equality—even in the one case where the husband claimed control (separate but equal)—far more than was the case in the 1970s. Several women also insisted that "women can do anything a man can do". The retired Local owner of a logging firm bragged to me that he had consistently employed a woman as a log truck driver, along with other attitudes indicative of what would have earlier been unusual respect for women.

> He remembers his own grade school teacher [a woman] with admiration and respect; he admires the very bright and successful wife of his brother (who didn't have a job, though perhaps he worked on their farm??); he treated me with complete respect despite our differences of [political] opinion; he had hired a woman (who also had an apparently affectionate nickname) as a log truck driver for years.
>
> *(notes, 2 June 2017)*

Although no mention was made of the team sports that were so central to masculinity in the 1970s, this may have been because of timing (the school year was ending) and/or sports' intimate links to the high school—the school being a context to which I had minimal access.[37] The recent, more inclusive attitudes reported by Anderson (2009) for college sportsmen in the US and UK may have reached or evolved in Bushler Bay as well. Sexuality, hierarchy and the arts were similarly not explicitly mentioned (which, given the brevity of this research, does not mean they weren't important).

A political harp string, which globally can be an important part of masculinity (discussed later) has become clearer among some Bushler Bay men. In the past, women took care of local politics and Public Employee men participated in their work lives, in governmental bureaucracies. Now, far fewer are employed, and many more who are, are women. Some of these elderly men now engage with politics, locally and reaching at least to the county level, suggesting a greater acceptance of hierarchy, as government bureaucracies are by their nature hierarchical. The fact that tourism, now so important in Bushler Bay, is more similar to bureaucratic work in its lack of a clear product (logs to the landing, loads to the mill) and 'time put in' rather than measurable output may contribute to a broader comfort with the hierarchy harp string.

Observations from Long Segar and Long Anai in 2019

In March 2019, I returned, first to Long Anai,[38] and then to Long Segar, East Kalimantan for two weeks (see Figure 4.1). During that time, I again conducted 'ethnography light' as well as a one-page survey administered to opportunity samples of 134 adults in Long Anai (total population: 435) and 143 in Long Segar (total population: 837). The survey in both communities repeated several questions that had been asked in a 1980 survey (Colfer 1985a, b), dealing with gender dynamics.[39] I had been communicating on Facebook with people in both communities, including scanning and posting old photos for them; they knew I was coming.

I was accompanied on the visit to Long Anai by my 95-year-old mother, my son (co-author Pierce)—both of whom had visited before—and his Javanese-American wife. We were welcomed, as usual, into the home of Pierce's father's previous wife. Although I had wronged her in 1980, she had long since forgiven me, partly because I had helped her and her family from time to time but also because of the norm that once one paid one's fine (as I had), the issue was finished (see Chapter 4). We were touched by the kindness and welcome these folks extended, particularly to my very fragile and inconvenient mother. They had insisted we bring her, even though her presence required them to buy a toilet, build a bed and borrow an easy chair, as well as helping her on and off multiple motorcycles.

I proceeded, after a week, to Long Segar, accompanied on the journey by Pierce's father (whose name has changed;[40] I'll call him Penjau) and two other elderly Kenyah men, one of whom I'd also known well before. Transport was provided free by one of the oil palm companies trying to gain more land in Long Segar. These Kenyah friends left me at the edge of the Telen River and returned to their home communities (feeling no longer welcome in Long Segar, because of conflicts over land). I was taken in by one of the sons of the family with whom

I'd lived in 1979–1980, and welcomed (in both communities, with more enthusiasm than in Bushler Bay).

As in Bushler Bay, I confronted very different worlds from what I had seen in earlier visits. The landscapes by which these two communities are now surrounded has changed utterly. Whereas previously there was a complex mosaic of primary and secondary forest, with agricultural fields interspersed, now the predominant feature of the landscapes is large-scale oil palm plantations.[41] In Long Segar, reportedly for three kilometres either side of the Telen River, land remains for the people to manage,[42] though the companies continue to seek access to more land. The same process is underway in Long Anai, with the addition that now mining companies have joined the fray and are also negotiating with local people to gain access.[43] Another new complication in Long Anai is the recent decision to move the Indonesian capital to their district.

In the following discussion, I return to the harp strings described in Chapter 4, and insofar as possible, update that discussion.

Expedition-making and masculinity

As noted earlier, a key harp string in Kenyah masculinity has been expedition-making. In the 1970s and before, men made their way to Malaysia and beyond; in more recent years, they have made extended trips throughout East Kalimantan. But Kenyah men are not now making such trips. The opportunities for land-clearing jobs, one of their longstanding skills, have evaporated as the forest has been cleared.

Additionally, because of external pressures on the land surrounding the two villages and resulting uncertainty about tenure, men are reluctant to leave. Many fear that in their absence, their family's land is more likely to be taken over. Although the Kenyah have traditionally considered a family's cleared land to belong equally to its men and women, outsiders (from the government, industry, even NGOs) tend to consider land to belong to men—insofar as ownership rights are granted to the Kenyah at all. Such rights are negotiated on a case-by-case basis.[44]

Leadership as a harp string

Although the issues being discussed differ somewhat, given the prominence of oil palm these days, the nature of leadership appears very similar to its traditional pattern in Long Segar. With the current *kepala adat* (customary leader) the son of Pelibut (the previous commoner headman), there remains a sub rosa concern about the *Paren-Panyen* (aristocrat-commoner) distinction,[45] with *Panyen* individuals arguing, for instance, that demonstrated ability to lead rather than one's ancestry should categorize one as *Paren*.

There also remains significant competition for recognition as leaders, among men of similar age. Three elderly men, whose competition was obvious in 1979, continue to compete for decision-making authority, power and resulting accolades in 2019. Two of these men are co-resident in Long Segar; the third, Penjau, lives elsewhere but has come together with leaders from Long Anai and Sungai Bawang (both daughter villages of Long Segar) in order to resolve a land issue that also pertains to the oil palm company that provided our transport.[46]

All these men had cleared land in Long Segar at one point, but the latter non-residents, had moved away decades ago. The question is, what should become of the land cleared by those no longer residing in the community? This was a bone of contention also in Long Ampung (the remote community from which the Long Segar population originated) in earlier times. Should non-residents retain rights to the land they cleared or should it return to the community's estate, available for use by those remaining?

This question becomes more complicated when oil palm companies' interest in gaining access to that land emerges. There is considerable scope for payoffs from the company to leaders and for manipulating existing community factionalism, thereby reducing local negotiating power. Formal Kenyah leaders are confronted with many temptations, a serious dilemma for some. Although such temptations are not new, as the amount of land available goes down, the adverse implications of leaders' possible collusion with companies goes up. Such troubles have beset both communities, with resulting community suspicion of men who lead (again, nothing new).

Considerable communication has long existed among leaders of different Kenyah villages, with reciprocal knowledge about each other, including cross-visits among them. Such communication is much easier now, with roads, cars, motorcycles, cell phones and Facebook—all of which are available to some residents of both communities. The fact that these new devices, as well as educational achievements, are more available to the young may at some stage have implications for the Kenyah preference for leadership by the elderly. However, elderly men appear to remain in control of major community decision-making at this stage (with easy opportunity for any who choose—man, woman, youth—to make their views known to these men).

Our 2019 survey asked each respondent whether he or she attended community meetings and whether they spoke up in them. Although we see slightly more gender differentiation in Long Segar (see Figure 7.1), in neither community was the difference statistically significant. But there are marked differences between the two villages, with 90% of women and 93% of men claiming to attend village meetings in Long Anai, and only 58% of women[47] and 76% of men in Long Segar reporting attending.

Figure 7.2 shows comparable data on the likelihood of women and men speaking up. Those in Long Segar are somewhat less likely to speak up in meetings than in Long Anai, perhaps reflecting the younger, more educated and more urbane population in the latter community. But again the gender difference is not significant in either community.

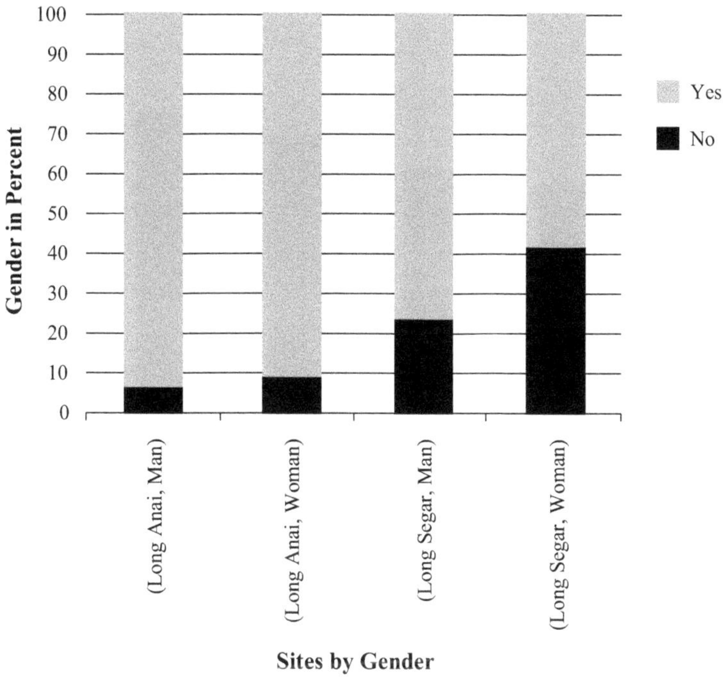

FIGURE 7.1 Self-reported attendance at community meetings, Long Anai and Long Segar, March 2019.

Provisioning as a harp string

As noted earlier, provisioning in general has been shared between Kenyah men and women, with men providing money and goods and women providing rice and other agricultural products. One would expect the fact that the percentage of people making rice fields has plummeted in Long Anai (95% of women and 93% of men reported not having made a 2018 field) would have a significant effect on the gendered distribution of labour. This is very different from Long Segar, where only 21% of women and 20% of men reported not making one. Despite the huge community difference in swiddening, the responses of men and women are not significantly different one from the other in either case. Women's upsurge in paid employment in the oil palm industry and involvement along with men in other agricultural endeavours in Long Anai appears to be substituting to some extent for their traditional rice production.

There is considerable evidence available on Facebook that the Kenyah continue to harvest some products from nature (particularly fish and some forest plants). The fact that they now buy much of their food (a big change from the past, when bought food in three surveys, 1979–1980, 1991 and 2001, ranged

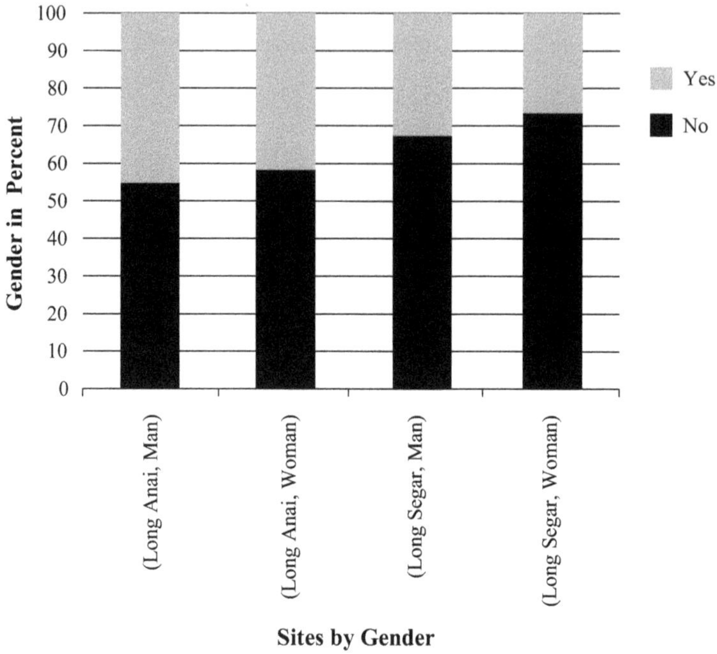

FIGURE 7.2 Self-reported willingness to speak up at community meetings, Long Anai and Long Segar, March 2019.

from 5–10% of their diet; Colfer and Soedjito 2003) derives partly from a simple lack of forest from which to harvest their traditional products. The reduction in hunting represents a reduction in men's provisioning roles.

Additionally, as land pressure has reduced the amount of available land, women have reduced the sizes of their fields and no longer sell surplus rice, which had been a ready source of cash and trade for them. Although men have typically had easier access to larger sums, through their expeditions and resulting wage labour, this is much less available now. Men complain that there are no land-clearing jobs.[48] That, combined with their reluctance to leave home due to concerns about land insecurity, means they are shifting their focus to smallholder oil palm and other commodities as potential sources of income. Families in both places have been developing a two-pronged approach: In Long Segar, the women continue to focus on rice for subsistence (sometimes supplemented by wage labour at the nearby oil palm plantations) and the men focus on *kebun* (gardens/orchards)—with each also contributing labour to the other's endeavours.

In Long Anai, both men and women in many families work for the oil palm industry,[49] also maintaining *kebun* for cash crops. Long Anai, only a three-hour drive from Samarinda, has been more involved in the sale of produce than Long Segar from its inception. Another motivation to develop *kebun* is the governmental perception that crops planted in rows are clearer evidence of use, necessary for any kind of external recognition (so far, still vague or non-existent) of rights to the land.[50] Finally, the development of *kebun* involves some tasks that are said to benefit from men's typically greater physical strength.

Insofar as Kenyah men's masculinity is closely linked to provisioning, the greater educational accomplishments of young Kenyah women may be seen as a potential threat. A number of Kenyah women, probably at least as many as Kenyah men, have office jobs with the government, NGOs or industry. A minority of husbands follow their wives. One of our relatives stayed home to look after their children. Although he would have preferred to work for wages, his main concern seemed not to be an attack on his masculinity, but rather his wish to demonstrate that he was a hard worker, a key pan-Kenyah value.

Our survey asked about how decisions relating to household income were made: Did you help decide how to spend household money last year? Alone? If not alone, with whom? (*'Iko' mepoh pekimet kompin pakai uang kem uman ca re? Tengen? Bek mpi tengen, ngan 'ee?'*).[51] The responses are shown in Figure 7.3, with no significant gender differentiation in either community, and not a great deal of variation from one community to the other. In both communities the most common response was that husbands and wives decided together.

Sharing was a strong value among Kenyah men (and women) in 1979–1980, and it continued strong in 2019. Walking along the street in Long Anai, I was

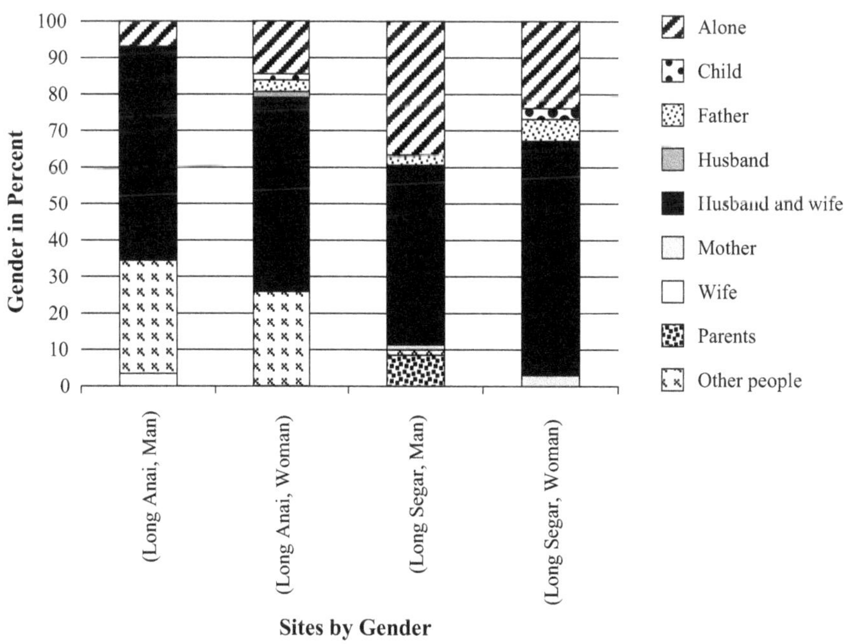

FIGURE 7.3 People with whom respondents made decisions about household money in Long Anai and Long Segar, March 2019.

offered some of any produce people might have in their backpacks or on their motorcycles. In Long Segar, the family I stayed with and the community at large organized get-togethers, where food was brought and prepared communally and shared among any who chose to come (visitors could also bring home banana leaf packets of food for housemates). On Facebook, people show what foods they have gathered or are eating; others express their desire to share it, a desire often recognized with invitations to come have some. Although I cannot confirm with current evidence, I suspect that gathered and home-grown products are more readily shared than those which are bought. That would certainly have been the case in the past. Some efforts are made to hide what people would rather not share, but any visible abundance is expected to be and usually is shared.

Protection, strength, courage and violence[52]

As noted in Chapter 4, the forest was considered a dangerous place, a place for men more than women. The man who was a midwife (see Chapter 4) had lost one leg to a wild pig; Dinis, one of our interviewers and my stepson, had a huge scar where a wild pig had attacked him and taken away part of his arm. Such encounters were not rare. The fact that the forest is basically gone means that men have lost an opportunity to display their strength and courage regularly.

In earlier visits, the Kenyah, both men and women, expressed concerns about *ayau*. *Ayau* is the term they used for headhunters in the past, but by 1980, they were using the same term for any violent non-Kenyah. The fact that there are now many more individuals from other ethnic groups living and working nearby has meant an increase, at least in Kenyah perceptions, of danger.

One proxy for people's feelings of danger is their willingness to sleep alone in their fields. Figure 7.4 shows women's and men's responses to this question in the two communities. Neither the gender differentiation nor the differentiation from one community to the other is significant. But this distribution of responses is markedly different from the 1980 results from Long Ampung and Long Segar, where ~84% of women in both communities reported staying alone at their fields (Colfer 1985a, p. 202).[53] *Ayau* were a source of fear at that time as well.

The Kenyah tend to have strong negative stereotypes of Bugis [an ethnic group originally from South Sulawesi], often noting their propensity for violence and aggressive use of knives. Most other ethnic groups (from Sulawesi, Nusa Tenggara Timur, Java) have more hierarchical views of male-female relations than do the Kenyah, with stronger narratives of female inferiority—thus suggesting greater *need* for male protection than had been the case in earlier years. As discussed in Chapter 5, interaction between ethnic groups can strengthen gender stereotyping as ethnic groups strive to emphasize the norms and roles of their particular group vis-à-vis those of others.

But Kenyah stereotypes of others are well matched by negative and fearful stereotypes about themselves. Pierce recounts here some stereotypes he's encountered in Boxes 7.2 and 7.3.

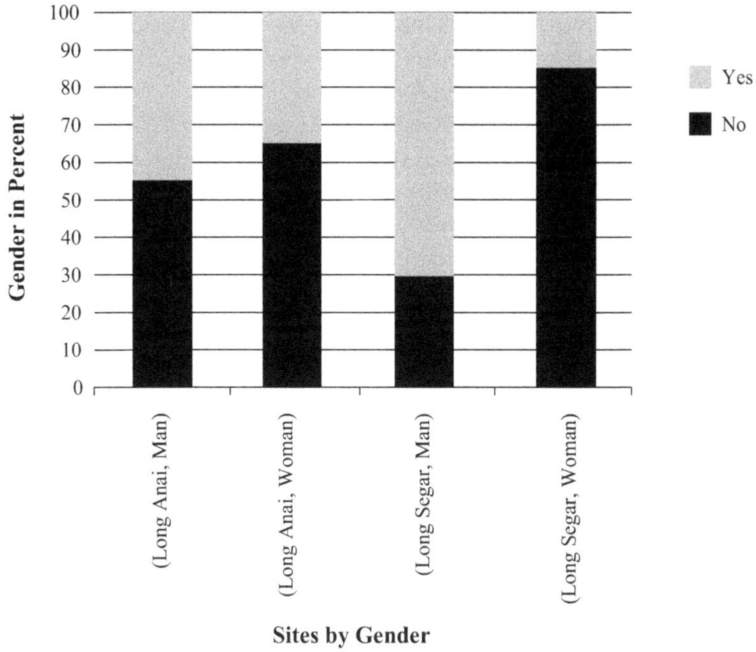

FIGURE 7.4 Willingness to spend the night at one's field without one's spouse, Long Anai and Long Segar, by gender, March 2019.

BOX 7.2 EXAMPLES OF UNFLATTERING AND FEARFUL STEREOTYPES ABOUT DAYAKS FROM OTHER ETHNIC GROUPS

Minangkabau: "Where is your spear and why are you wearing clothes?" a woman from West Sumatra said jokingly to Pierce. The same woman, Pierce's friend, told him, "You are protected; I see spirits that protect you, that are always with you".

Ambonese: "Ah Dayak, your magic is almost as strong as ours. In my village, our *Dukun* [magic healer] can fly on a leaf; can they do that in your village?" said an Ambonese man (also reputedly a killer-for-hire in Bogor, Indonesia, 1998).

Javanese: "Don't go to Kalimantan, because you will come out with a penis on your forehead if you ever try and leave a Dayak woman. Most people who go there never leave". Also "Dayaks have powerful love magic".

Australian-Indonesian: "I don't believe in magic, but I've seen a lot of friends of mine be mind-controlled (*santet*) to fall in love with Dayak women. A good friend of mine loved his wife, was the perfect family man, had pictures of all his kids in his office and one day he just left . . . left his family, his job and started living with this woman. We couldn't believe it, but I saw it happen and there were many other stories like that" (from friend who had long worked in mining in Kalimantan).[54]

Assorted Indonesian ethnicities: "Dayaks can send a sword that flies through the air and it will kill you if you are of Madurese blood". This sentiment was heard repeatedly after the 1996–1997 inter-ethnic violence in West Kalimantan between Dayaks and in-migrants (see Peluso 2008 for a good summary of this violence). "Dayaks can smell a Madurese. They'll come on a bus and pick out the Madurese by their smell".

Additionally, my well-educated and intelligent Balinese hostess in 1979 expressed fear when told that I was going to East Kalimantan. She believed that Dayak women were very white and thus particularly beautiful, and that Dayaks had tails.

Here, to better capture Kenyah men's perceptions of manhood, I have enlisted further aid from Pierce, my son, who has spent almost half his life in Indonesia. Borrowing the eyes of a man in his 30s seemed a useful complement to our survey results. Although much of Pierce's life has been spent in urban Indonesia (Jakarta, Bogor and more recently, Denpasar, Bali), he also spent three years in rural West Sumatra (from ages three to five), another year in remote West Kalimantan (age 11), another year as a construction manager on the rural island of Matak in Riau Province (2011–1012) and has periodically made family visits to rural East Kalimantan.

He has had moments of some fame in Indonesia (due to a brief advertising career, then brief involvement in televised cage fighting and other martial arts). Part of his adult life, he focused on his artistic talents, and then shifted to entrepreneurial activity in construction in the US, then recently to several businesses in Bali. He has recently again plucked his artistic harp string.

As a 'people person', whom others usually enjoy, I enlisted his help in recounting what he'd learned about current Kenyah masculinities as shown by his half-brothers and others in Long Anai. Pierce's size (6'2"), strength and fighting skills (proven in a video of his nationally televised cage fight from 2003) were obviously admired in this community, and he was entertained nightly with stories of men's lives in Long Anai. We include several such tales here, reflective of masculine actions and perspectives.

BOX 7.3 THE FEARS OF OTHERS ENCOUNTERING THE KENYAH

In Long Anai, Taman, a Tunjung Dayak married to Pierce's Kenyah half-sister, told a story of a Javanese palm oil plantation worker who avoided others, never speaking to anyone. He ate alone and always looked frightened. After a couple of weeks, Taman went to him and introduced himself, to break the ice. He felt sorry for the guy, without friends. The man nervously gave him his name and immediately and fearfully said the first thing on his mind: "In my village, they say Dayaks eat people, but I haven't seen you eat anyone". Taman described the man's reply, with incredulity and laughter.

"You thought that we eat people and you still came here?!" Smiling, Taman said, "I told him that we did not eat people and that that was generations ago and rarely happened, that none of the Dayaks in my lifetime have eaten people". [He judiciously excluded the 1997 Dayak-Madurese conflict in West Kalimantan.] Taman patiently explained his Tunjung and Kenyah life philosophy and culture. The man's fears eased over time, allowing him eventually to feel comfortable surrounded by Dayaks.

Several stories were recounted about groups of Kenyah men banding together to confront outsiders—an event the frequency of which has only increased in recent years. Physical strength shows up as an admired harp string. In one case, the men of Long Anai invited a group they referred to as the 'Dayak mafia' (a group of hard drinking, womanizing, drug-using urban Dayaks)[55] to help them in a conflict with an oil palm company. The Dayak mafia was reputed to be 'skilled at the table' (able to bluff, pound the table, threaten). However, in this case, that approach failed, and the group decided to resort to violence.[56] With spears, bush knives and a few guns in hand, they began a two-kilometre hike to company headquarters. But the Dayak mafia, composed of fat, slow 'city folk', was reported not to have the stamina even for the walk, giving up halfway there. The Kenyah recounted this story with great disdain for the physical state of these visitors. One said, "How are they going to fight if they can't even get to the battlefield?!"

In another case, which documents the kinds of trickery to which the people have been subjected, violence proved a harp string worth plucking (though apparently not successfully). A huge Korean mining company was not hiring local people as promised in their original agreement, which granted the company access to land. The people came together en masse to complain and demonstrate. The company backed off and agreed to improve their practices. The head of the district agreed to put in a two-week apprentice programme to train local people. He is reputed to have received $700,000 from the company to implement the programme. However, nothing happened, and the mine kept operating. Such actions by companies and government officials have led to bad blood with the communities and retribution in the form of reciprocal community trickery, as also reported elsewhere. Legal means are not helpful: judges can be bought off by the wealthier companies. People feel their only recourse is violence, usually by men (notes, 19 March 2019), though women are sometimes included.

Stories of hunting praised men's skills and derring-do (see Box 7.4). Interestingly, Pierce, a strong and demonstrably courageous man himself, was told more stories of this ilk than I, in this short visit. This story portrays other masculine harp strings reminiscent of masculinities among scientists: 'being good at speaking' (evident in the telling of these tales) and having knowledge and analytical skills.

BOX 7.4 BILUNG, AN EXCELLENT HUNTER FROM LONG ANAI, MARCH 2019

Bilung has a reputation as a good hunter. He shows me a picture and the skin of a six-metre python he killed. He was able to sneak up on the python, pre-occupied eating a baby pig, and kill it by stabbing its head into the ground with a spear. The baby pig tried to run away, but Bilung was able to catch and kill it also, with a bush knife.

On another occasion, he told of killing a honey bear as night was falling. Its mate came and chased him; he tried to fight it with a stick but realized that fighting at night was going to be even more dangerous. He couldn't run up a tree because the honey bear would follow and trap him there, so he ran away . . . but he really wanted to do battle with the bear. He said to himself, "I will retreat now, but you will see tomorrow. I will come for you and we will fight". He returned the next day, finding the bear in the same clearing where he had killed its mate. The bear, in fear, ran away.

As he recounted this tale, he explained the advantages of his technique over that of one of his agemates. He broke down the elements of strategy and skill necessary to hunt well, the art of reading tracks and body and weapon positioning. He called himself a tactician. His friend, on the other hand, has great courage, he said, but is not sufficiently tactical. "You watch an animal's movements and patterns and study them. Then you can hunt well". He then told a story of his friend's incorrect positioning, which nearly got him killed. "[His friend] is reckless; this is very dangerous. You have to be tactical and always secure a superior position before you attack".

Once, for instance, Bilung followed two adult pigs with two little ones deep into the forest. He saw where they were headed and got on a log below which he predicted they would pass. Having only a machete, he threw it at the big male, hoping to stab it cleanly, but the machete bounced off and the adult pig ran. Bilung jumped onto the piglets, catching one, wrapping it with one arm and his legs, and grabbing the other with his other arm. Unable to kill them both because his machete was out of reach and with no hand free, he let one piglet go, and managed to kill the other. He was in great danger, as an adult pig can kill a grown man easily with its tusks or by its bite, let alone a male and female pig protecting their young.

Bilung was also one of the few willing to hunt crocodiles with spear and gun. He would track them by their eyes which protrude from the river and shine in the night. He used a barbed spear, striking through the mouth and then shooting them in the head.

One of Bilung's cousins, Dinis, was also an excellent hunter. He would sometimes take his makeshift rifle with only three bullets into the forest and stay for two to three days alone, hunting wild pigs. Since bullets are expensive, and no one could afford modern firearms, he used a gun that required him to jam a bullet into the barrel with a rock each time he reloaded. I was with him on such a hunt in 1999 and feared the whole while that the bullet would explode in his face, showering us with shrapnel. Thankfully this did not happen. (Pierce's notes, 19 March 2019)[57]

These two men (both *Paren*) have invested strongly in the strength, skills and courage harp strings; the teller has plucked the articulateness harp string. Not all Kenyah men choose strength, courage or articulateness.

Another story from Pierce, however, reveals womanly courage, a lack of male protectiveness and a willingness by men to acknowledge fear.

BOX 7.5 A KENYAH WOMAN'S COURAGE

In 2009, I'd gone fishing with my brother-in-law (the above-mentioned Tunjung Dayak) and his Tunjung family also visiting Long Anai. These Tunjung had little experience with fishing but were famous instead for climbing the tall, majestic honey trees (*Koompassia excelsa*), the latter widely recognized as requiring great courage. A group of Kenyah men, excellent fishers, agreed to try to teach them all how to catch fish with their hands. On this day, the Kenyah men caught many fish, whereas the Tunjung were coming up empty-handed.

Halfway through the day, I heard a high-pitched scream and turning around, saw a thin stripling of a tree—from which dangled my brother-in-law, swaying about a metre above the water, legs splayed and shouting, "Snake!!! Snake!!!" The other Tunjung men had scattered to the river bank like buckshot. Half laughing, half fearful, I ran to shore, picked up a machete and a stick to pin the snake's head against the tree. I yelled out, "Do you see it?". He answered, abject terror clear on his face, "I see its head. It's right there. I see its head!" I waded back into the water as all the Tunjung stood huddled together on the shore, unwilling to re-enter the water. As I got closer, I saw the snake's head at the base of the tree. Just as I was about to pin it with the stick, the head disappeared. I waited a moment and then, like the Tunjung, raced back to shore, also not wanting to get bitten by a snake!

Just then, two Kenyah women passed by with rattan backpacks three quarters full of fish. . . . about 20 times the amount that we had caught, our sole saviors being Bilung and his brother. One of the women yelled out, "What's the problem?"

"There's a snake under that tree", I said, "but it disappeared".

"Oh, just a snake". She turned around, pulled her machete from its casing, cut a forked branch, hacked off the two ends and waded through the water to the tree. We all watched, nervously. In a matter of five seconds she'd pinned the snake's head, cut it off, grabbed the body and flung it into the forest. She walked off laughing at our ineptitude and cowardice, having out-fished all nine of us. When I reminded her of this outing in March 2019, she just laughed dismissively.

On another occasion, another woman, known for her courage, held a python by its throat and allowed it to wrap itself around her arm, wondering how hard it could squeeze. When her arm began to turn blue, she smashed in its teeth, and unwrapped it from its tail (reportedly the 'proper way' to remove a python). Her husband had fled. (Pierce's notes, 19 March 2019)

Sexuality

Getting a proper sense of people's attitudes toward masculinity and sexuality may be more sensitive and thus iffier on a short visit than most topics. Kenyah attitudes appear to remain quite open about sexuality, however, and important to men. In one of the homes where I stayed, one adult son was openly sleeping with his fiancée, without any evidence of disapproval. However, as would have happened in earlier times, young people were discouraged from making love before marriage. At an engagement party we attended in Long Anai, four elderly men gave speeches of advice to the young couple, and all reminded them that they should not be making love yet (along with pleas for understanding of their cultural difference [one was Kenyah and one Batak], and for love and problem solving between them). Interestingly, in contrast to weddings attended in Long Segar in 1979–1980, no mention was made of the marital sex roles American missionaries had then deemed appropriate for men and women.

There continued to be community approval of marriage as the most appropriate life path for both men and women. In Long Segar, some concern was expressed by elders that a fair number of young people were not marrying. I encountered two unmarried men and two unmarried women in their 20s and 30s, all reluctant to marry, though I don't know why.

The greater amounts of interaction in both areas with individuals from other ethnic groups has linked sexuality and violence in a way that did not appear to be the case in the past. In one example in Long Segar, a man from another ethnic group and a Kenyah women wanted to marry; he brought her presents as a sign of his commitment. She then changed her mind but did not give back the presents. The man did not find this acceptable and kept harassing her about the issue. A group of Kenyah men, unhappy with his insistence, went *en masse* with weapons to find the man and beat him up.[58] They were stopped by the timely arrival of the police in the other community, who were able to negotiate a peaceful solution.

A similar case occurred in Long Anai. A non-Kenyah man became friends with a Kenyah man and his wife, eventually resulting in her disappearance. It was not clear whether the other man had persuaded her to leave her husband or had abducted her. Her husband was not inclined to pursue the matter, but other community members felt that it was important to find out whether she'd gone willingly or not. If she had indeed been abducted, they felt some compulsion to demonstrate more broadly that the Kenyah would not put up with such behaviour. They recognized that one of the few powers they have is other ethnic groups' fear of them.

There were also tales told by older men. One man of 72 (well known in his younger years for sexual peccadilloes) recounted the jealousy of his considerably younger wife. She had gotten so mad one time that she'd bitten a chunk out of his thigh, a story told far and wide (and with amusement). Another Long Anai man in his early 70s had taken his wife to Kenyah court for her jealousy (see Box 7.6).

BOX 7.6 SEXUAL JEALOUSY AMONG THE ELDERLY

On a long trip, in a car with three elderly, politically active men: One man (in his 60s) had just had a fight with his wife. She was mad at him about an incident from three years earlier: There had been a pretty woman from another village who liked him and sought him out (it wasn't clear whether he responded or not), and his wife got very jealous. At the time, people came together to hear this problem, and they asked both the man and the other woman if they'd been sleeping together. They both denied it. At the time, his wife was fined for being so jealous. Recently in the purported co-respondent's household, a new baby had been born, who was the man's 'grandchild' (not necessarily a direct lineal descendant). He went to see the baby, raising his wife's ire again. During the argument, he reported grabbing his genitals and saying to her, "*This* is what makes you so jealous!". The other two men listening to this story nodded in sympathy. There seemed to be general agreement with this interpretation; all also agreed that getting jealous is a bad thing. One of the other men had commented earlier that he liked to look at pretty women, but his body was weak now and he couldn't do anything about it.

A final example comes from a recently widowed man of around 80 in Long Segar. He had fallen in love with a young woman who was polite but not interested in him romantically. His agemates berated him to me, feeling that he should find a woman closer to his own age. They also disapproved of his continuing to pursue the younger woman, when they felt it was clear she didn't want him—though she had never exactly said so. One elder said "Women are polite. They're not going to say, 'I don't like you.' They'll make up excuses". This elder went on to say, "I'd never pursue a woman who wasn't interested in me. There are plenty of women, and it's easy for a man to find another woman". He clearly thought it was easier for a man to find a wife than for a woman to find a husband (which indeed seems to be true).

The interesting thing about these examples comes partly from a comparison with the situation encountered in Bushler Bay, where sexuality—so important in the 1970s—did not come up in my interviews at all.[59]

The arts

Although excelling at the arts is not an avenue open to all men—since most consider artistic ability to be a special talent—it is a legitimate avenue for those who are able. As noted earlier, Long Anai was declared by the Indonesian government as a *Desa Kebudayaan* (a Cultural Village). What that has meant is that there have been governmental efforts and incentives to keep Kenyah arts alive there. These include wood carving, music, dancing, boatmaking and comedy.

There are regular Kenyah dance performances shown on Facebook, with men as active in these performances as women. Kenyah men's dancing involves a great deal of alternately slow and fast movements, usually with bent knees and exaggerated eye movements. The male dancers still imitate both hunters and the hunted, wielding shields and bush knives and wearing animal skins, with headdresses made of hornbill feathers.[60] They demonstrate their agility, grace and strength.

I was not going to be able to visit one man, Gun, who had been a child in the household where I lived in 1979–1980. We were both disappointed, so one of his relatives called him on the cell phone, while listening to another man, Rinto, play the haunting music of the *sampe'* (a Kenyah 'guitar'). Both these men were recognized as excellent performers, and while Rinto provided the music, Gun performed a Kenyah dance for us, visible on the tiny cell phone.

Rinto was interested in all the arts and felt that he might have inherited the spirit of Pelibut, the Long Segar headman who'd been talented in so many ways, artistic and otherwise. As before, the Long Segar community selects a young man who provides leadership to the youth. He organizes events, encourages them to dance and participate in sports and community life. Rinto is that leader. When my Long Segar family wanted to dress me up in Long Segar traditional attire, he was recognized as something of an expert in how to put all the pieces together (headgear, beads, hat, vest, skirt and hornbill feather hand decorations) appropriately. He bewailed the loss of beadlore (Colfer and Pelibut 2001), cultural meanings of specific designs and other traditional knowledge that has been or is being lost.

The arts of comedy and storytelling also remain alive and well. Pierce's older half brother, Dinis, can hold a Long Anai crowd enthralled for hours by the tales he tells, a talent that has long been admired—closely linked with the harp string for public speaking. His ability to keep an audience laughing until the wee hours of the morning ensures his popularity. Pierce was told, "Aside from being a courageous fighter, great hunter and fisherman, he is the best entertainer in the village". Another relative, Pierce's cousin, Libang, like Rinto in Long Segar, plays the *sampe'*. Libang is sent to various cultural expositions and thereby brings pride to the village by promoting and preserving traditional Dayak arts.

Kenyah summary

Men's expedition-making harp string has largely disappeared and their provisioning harp string has, as a result, changed dramatically as well. Men's perceived need for strength, courage and potential violence as protectors of family and village lands has re-emerged, reactivating harp strings that were more dominant during headhunting days of old.

Men's emphases on sexuality, rhetoric and politics and leadership remain strong, as does the arts harp string. People, including men, still share any visible abundance within their group. Land tenure issues have become still more worrying as people continue to lose access to their traditional areas. One enduring feature is the comparative equity and mutual respect evidenced in relations between men and women in daily life.

Conclusions

From a people-and-forests standpoint, these contexts represent three very different trends.

- The forests of upstate New York, where I now live, are growing back as the population empties out and people move to cities in search of economic opportunities. Corn fields and dairy farms revert to forests, adversely affected more by an over-abundance of deer than by human activity.
- The forests of Bushler Bay have been maintained by taking away local livelihoods (or at least requiring a major shift in emphasis), while protecting the habitat for endangered species—with major demographic and cultural shifts a result.
- The forests of Long Segar and Long Anai have been decimated, replaced by hundreds of hectares of oil palm with only small patches of regrowing forest, tiny swiddens and smallholder plots (increasingly mostly oil palm) remaining near rivers—a process to which local people have adapted amazingly resiliently alongside their mixed feelings of loss and gratitude.

In my corner of New York State and in Bushler Bay the polarized traditional imaginings of masculinity and femininity show signs of moderating, as a more muted version of gender difference—consistent with changing global norms—grows. In Kalimantan, where gender was muted in the 1970s, stronger differentiation, emphasizing and building on the greater physical strength of men, is encouraged by the growing insecurity of land tenure and the 'invasion' of other ethnic groups, some more powerful, whose systems are characterized by narratives of women's inferiority. Despite these worrying trends, relations between men and women continue to be comparatively equitable there.[61]

Notes

1 Pierce, my Kenyah-American son, participated in the fieldwork, collecting and recording stories in Long Anai. Erlita performed the statistical analyses in the third section of this chapter and prepared the figures.
2 E.g., the Convention on the Elimination of all forms of Discrimination Against Women (CEDAW), ratified by 189 states in 1981; The International Labour Organization (ILO) Policy on Gender Equality and Mainstreaming since 1999 (www.ilo.org/gender/lang–en/index.htm, accessed 1 May 2019); the Millennium Development Goals from 2000, and the Sustainable Development Goals since 2015; the World Health Organization (WHO) global plan of action to address violence against women and children (www.who.int/news-room/fact-sheets/detail/violence-against-women; accessed 1 May 2019); the populist 'me-too' movement; the Global Environment Facility (GEF) Policy on Gender Equality in 2017 (www.thegef.org/news/new-policy-gender-equality-gef; accessed 1 May 2019), and more.
3 "No person in the United States shall, on the basis of sex, be excluded from participation in, be denied the benefits of, or be subjected to discrimination under any education programme or activity receiving Federal financial assistance" (https://www2.ed.gov/about/offices/list/ocr/docs/tix_dis.html, accessed 2 May 2019).
4 My father, when asked to extend his contract to continue teaching as a visiting scholar in Dubai in the late 1980s, negotiated the right to arrive late for the term, so that he could watch the Superbowl in the US.
5 For the first 25 years of our marriage, we had household help (discussed in Chapter 6), so neither of us performed these tasks.
6 Prior to COVID-19.
7 The amounts paid for this household labour are nowhere near as inequitable as in international contexts. Many here pay \$20–\$40/hour. If such employers make \$100,000/year, their hourly rate would be \$47/hour before taxes. Even if incomes are higher—which some are—the divergence from the wages they pay workers is still far less than in the Indonesian examples in Chapter 6.
8 These two are 'Third Culture Kids' (Pollock and Reken 2001), as am I.
9 Remembering that the rural New York example is a ten-minute drive from Cornell University, that my social circle reflects that proximity, and is quite different from the surrounding countryside where Donald Trump and his political and social attitudes dominate.
10 Given the smaller number of older American men compared to older American women, demographically, combined with American informal age-grading (Colfer 1977) Hatch's (2000) comments may have relevance:

> Other studies . . . provide modest support for the thesis that women have greater economic and political power in low sex ratio societies (low numbers of men relative to women) than in high sex ratio societies.
>
> *(p. 118)*

11 Colfer, Cerveny, and Hummel (2019) and Colfer (2018) provide fuller details of these and other results from this study.
12 I did not see the many political signs expressing support for President Trump that I regularly see in rural upstate New York, and only one man expressed his commitment to Trump and his politics to me, face to face.

13 "In-migration that occurs in a place because people are drawn to its natural and social features" (Charnley, McLain, and Donoghue 2008, p. 744).
14 All names are pseudonyms.
15 This refers to the Local–Public Employee social structural distinction (see Chapter 3).
16 Colfer and Colfer (1979).
17 'Pecker' is American slang for penis.
18 See Buttolph et al. (2006) for an assessment of changes in this community from 1994–2003.
19 Protection of the spotted owl, and later the marbled murrelet, was the immediate spur for a great deal of pro-environmental action. A few months before I arrived in 2017, a restaurant called Loggers Landing had closed. It had advertised "spotted owl soup" on the menu; in earlier times, pro-logger folk sported T-shirts that said, "Save a Logger; Kill a Spotted Owl".
20 Factual correction: To my knowledge there had been no mill in or near Bushler Bay for at least half a century.
21 In most of this book, I include Hood View in Bushler Bay, as most social patterns were similar.
22 Based on interviews with six mostly elderly individuals knowledgeable about the communities.
23 Proximity, pre-existing rights to forests, dependence on forests, poverty, local knowledge, culture/forest link and power deficit.
24 This pebble distribution tool was conducted in three mixed-gender group discussions, and with six women and three men individually. The emphasis on the elderly is an artefact of the availability of interviewees, but also reflects their demographic predominance.
25 Salal, brush, etc.; shellfish/fish; berries; firewood, poles, etc.; Christmas trees, wreaths, etc.; mushrooms; and animals/game. We interviewed two groups (aged >50 in Bushler Bay) and eight individuals, totaling 22 respondents. Half the individual respondents were <50; 14 were women, eight men.
26 Jackson (2001) concludes that performances of masculinity are age-specific, using examples from Kerala, India (pp. 5–6). See also e.g., Bribiescas 2016; Hatch 2000; Varley and Blasco 2001, on men and aging, from very different perspectives.
27 Tom was recommended to me by a long-resident, educated woman friend whom I'd known in our youths, as an interesting and opinionated man whom she liked but with whom she also often disagreed.
28 Cornwall (2016) notes that "Standing like a man [in Sierra Leone] . . . is also about being able to enable women's consumption of consumer goods, in a context where romance and finance are intimately interwoven" (p. 19); or Hayns 2016: "One dominating conception of masculinity in Morocco (and, of course, elsewhere) is that men—if they are indeed men—must make money and with it provide women, usually their wife or girlfriend, with material subsistence" (p. 105).
29 This symbolism is nothing new or unique to Bushler Bay: Remember my first husband's desire for a pickup truck in 1972 or watch the Dodge Ram ads on TV today. There is an F150 sandwich on the menu at the newer, more yuppy café in Bushler Bay. When my husband and I take breakfast at a rural diner in upstate New York (the equivalent of Bushler Bay's Local lifestyle), the pickups out-number the sedans many times over, with Subarus and Priuses identifying their drivers as outsiders, alien. That these trucks are American-made is not irrelevant either in Trump's America (though Trump's politics are not as popular in rural western Washington as in upstate New York).
30 I discovered Dick, through a friend of a friend in Portland, Oregon, who had used his construction services for her home on nearby Whidby Island. I imagined this would provide a new and different entrée to the community.
31 I'd known Harry marginally in the 1970s, but he was recommended by a Local woman I knew better, because of his knowledge of the brush industry (see https://products.kitsapsun.com/archive/2000/11-23/0002_brush_picking_-_the_silent_indust.html).
32 These folks identified as non-American do not live locally. They are reportedly brought in, typically contracted by bigger businesses, by the busload. Davis et al. (2020) note

that the USFS often hires immigrants for labour-intensive tasks in the American West. Charnley et al. (2018) found that

> on Washington's Olympic Peninsula—a focal point for the Northwest's floral greens industry—the harvester workforce was originally Euro-American, but shifted in the late 1970s and early 1980s to being dominated by refugees from Southeast Asia, then shifted again in the late 1980s to become dominated by immigrants from Mexico and Central America (McLain and Lynch 2010). Asians are also active participants in commercial wild mushroom harvesting, particularly matsutake (*Tricholoma magnivelare*).
>
> *(Charnley et al. 2018, p. 827)*

33 One unemployed Local man of ~50, working as a volunteer fireman, told of the patch of mushrooms he had found and delighted in keeping secret (notes, 30 May 2017). Another, a retired Public Employee who avoided contact with the community, regretted the loss of the forest in which he had habitually hunted mushrooms. It had disappeared when a private company logged the land adjacent to his property in 2016 (notes, 26 May 2017).

34 Or the situation described by Reed (2003) where "some women who were interviewed [on nearby Vancouver Island in 1997] chose not to risk challenging the 'manliness' of their partner by taking a job" (p. 384).

35 He did not give the impression, however, of having 'lost' his masculinity in the way described by Vera-Sanso (2016):

> By taking the long view, we can see that in South India men typically follow a trajectory in which they attempt to achieve, and sustain, an adult masculinity, based on a socially enforced role as provider and head of the family, eventually declining into a feminized status as an aged dependent.
>
> *(pp. 82–83)*

36 Although not stressed explicitly in the cases presented, courage came up in other interviews, e.g., one with a long-time firefighter produced the following response, among other courage-related remarks: "I asked if he felt fear in the forest. He said 'no, I've been chased by a moose, and been near grizzlies, but such experiences are part of the mystique'" (notes, 30 May 2017).

37 Fearing Cornell's IRB (Internal Review Board), I failed to request permission to engage with high school students and was thus precluded from interviewing those under 18 years of age, an error of judgment on my part. I also encountered greater reluctance in the school than in any other institution, perhaps due to rumours and the displeasure of one or two community members with the contents of my earlier studies, though legalities about dealing with minors may also have been the main issue.

38 Long Anai is a daughter village of Long Segar, located a few hours south of Samarinda. Long Segar residents began establishing it after the 1983 El Niño.

39 The Indonesian government required that an Indonesian conduct the interviews. In Long Anai, Tamen Loren and Pesuyang (a Kenyah man and woman in their 40s) conducted the interviews; in Long Segar, Dorkas and Erni (two Kenyah women in their 20s) did so.

40 The Kenyah change their names regularly, with all of them adopting a new name beginning with 'P' or 'Pe' when they become grandparents.

41 "Kalimantan's oil palm plantations have expanded rapidly over the last decades, covering a total area of 13,000 km^2 (in 1,073 villages) in 2000, tripling to 40,000 km^2 (in 1,980 villages) in 2015" (Santika et al. 2019, p. 108).

42 This land is part of Indonesia's Forest Estate, and thus the government has the legal right to manage it; however, local people also consider the land to be theirs, according to their traditional land tenure system (land belongs to the person who first cleared the primary forest; see Marfo et al. 2010, for discussion of the Long Segar and Sitiung tenure situations).

43 A Long Anai resident posted a video on Facebook (December 2019) of Long Anai men armed with bush knives and spears angrily demonstrating at a new mining operation on land the men considered their own. The Indonesian military was in evidence as well.

44 Siscawati (2020) also found that men make fewer expeditions in Lampung Province, though she attributes this to the effects of the government's tenure reform efforts—specifically, *Hutan Kemasyarakatan* (HKm, community-managed forests) and *Hutan Tanaman Rakyat* (HTR, community-based plantation forests), which have strengthened tenure security, exactly the opposite of trends in Long Segar. These programmes, not implemented in Long Segar, appear also to have enhanced women's empowerment and involvement in formal forest management in Lampung. See Colfer, Monterroso, and Ihalainen (2020, in press) or Elmhirst, Siscawati, and Colfer (2016) for discussions of the impacts of these landscape changes, new tenure regimes and new powerful actors on women in Long Segar and Long Anai.

45 See Chapter 4.

46 The conclusions of Santika et al. (2019) appear to apply to our study communities as well:

> We found an overall increase in basic, physical and financial indicators of well-being between 2000 and 2014, both in villages with oil palm plantation developments and those without such developments across Kalimantan between 2000 and 2014 . . . Conversely, there was an overall decline in social and environmental measures of well-being . . .
>
> (p. 111)

Ironically, *those with oil palm development showed slower increases in the first measures and faster declines in the latter measures.*

47 In the 1980 survey, in both remote Long Ampung and Long Segar, 56% of the women reported attending meetings (from 'rarely' to 'always'; 'sometimes' being the mode, 40% in Long Ampung and 43% in Long Segar). Although I did not ask then whether the women spoke at community meetings, I observed that they rarely did.

48 Evidence supporting this local observation is available from remote sensing. The rate of forest loss declined precipitously after 2016, and new plantation development (thus land clearing) began declining after 2012 (Gaveau et al. 2018).

49 One survey question, "Did anyone tell you what to do today? If yes, who told you?" (*A'un dulu dia' ca cuk Iko' uyan inu tau ini*? *Bek ya, ee' ya' cu' iko'*?) was intended to ascertain who was telling whom what to do, but in the Long Anai case is probably more of a proxy for wage employment. There, this was typically interpreted to refer to what happened at work; in Long Segar it was interpreted more broadly, including as originally intended. One hundred percent of Long Anai women respondents indicated they'd been told what to do that day by 'folks working together'; for men, 78%. In Long Segar, where there was a greater likelihood of the traditional shared work parties (Indonesian: *gotong royong*; Kenyah: *pemong gayeng*) and fewer individuals formally employed, the figures were 16% and 26%, respectively. For comparison, a similar question was asked in Long Ampung and Long Segar in 1980: "Of the 15% in Long Ampung and the 20% in Long Segar who responded affirmatively, almost none mentioned their husband or any other man as the person who had assigned the task" (Colfer 1985a, p. 199). In these new data, men and women in both communities commonly told each other what to do with no significant gender differentiation (2019 survey).

50 The question of land rights in Indonesia is extraordinarily complex. For decades, the claims of rural people living in forests were not recognized at all. In recent years, the central government has developed several social forestry schemes, which are experimentally implemented in some areas. This CIFOR video (Securing Customary Rights to Forests in Maluku Province, Indonesia) outlines some encouraging discussions recently underway about forest tenure in Maluku, where communities want formalization of their traditional systems and full management rights rather than a shared system. Or see

Fisher et al. (2017) for an analysis of a lengthy attempt to bring together the interests of a traditional community in South Sulawesi with those of formal forest managers, with varying success.

51 This question in 1980 was "Did you help decide how to spend the money you got from selling rice last year?" in Long Segar and "Did you help decide how to use last year's rice?" in Long Ampung (where money had little use). In both these communities, 64% of the women reported helping decide.

52 This section particularly has benefitted from the insights of Andrew Balan Pierce (half Kenyah, aged 37). His interactions with the Kenyah in Long Anai brought forth much more discussion about courage, strength and forest-based skills than did my own interactions.

53 Sadly, that study involved no men respondents.

54 See Manurung (2019), for similar stories about Sumatra's Orang Rimba.

55 The existence of such urban Dayaks represents another difference from the 1980 context, when almost no Dayaks could be considered urban.

56 Resorting to such threatened violence was not new. The people of Long Anai said they "threatened the HTI [industrial timber concession] workers with blowpipes and drove them away four months ago" (journal, 28 March 1996). The December 2019 Facebook video of such a confrontation has already been mentioned.

57 In Long Segar, Colfer's fictive brother's son had gone hunting in the forest. Thinking to have a stronger shot, he had rammed multiple bullets down the barrel of his gun, and it had indeed exploded in his face, blinding him for life (notes, 16 March 2019).

58 A somewhat similar conflict was the original impetus of the deadly Madurese-Dayak conflict in West Kalimantan:

> On 29 December 1996, at a pop music concert in Ledo, Sambas District, West Kalimantan, two Dayak youths were stabbed by a group of Madurese seeking revenge for being humiliated at a previous concert after 'bothering' a Dayak girl.
>
> *(Peluso 2008, p. 49)*

59 A quick search of my 2017 notes on Bushler Bay, for 'sex' turned up no entries and for 'love', 50 entries, the vast majority of which were love of the environment. Ten percent were love of animals, 8% love of work, with another 10% scattered among four topics. The only love relating to humans was mothers' love for their children (4%). All this is interesting in light of recent articles on a 'sex recession' in the US.

60 Roberts (2011) describes the attitudes of three Native Americans from the Plains about dancing:

> These three men . . . can be said to represent three different personal perspectives about competition, war, and masculinity in powwows. Ira respects competition in powwows, and participates in competitive powwows, but makes a personal choice not to enter the contests. Jim is drawn mainly to the spiritual aspects of dancing and also is grateful for the money he might win in competitions. Doug . . . competes *because* of the communally spiritual rewards of dancing. He . . . describes himself significantly as *'a modern day warrior'*.
>
> *(p. 155, italics in original)*

See Spiller (2010) for Sundanese men's attitudes toward and performance of dance.

61 Siscawati's (2020) observation that gender norms had not changed much with the advent of formal social forestry programmes in Lampung Province, in this case remaining *in*equitable, also speaks to the deep and enduring nature of gender relations.

References

Anderson, Eric. 2009. *Inclusive Masculinity.* New York: Routledge.

Bari, Judi. 1994. *Timber Wars.* Monroe, ME: Common Courage Press.

Bribiescas, Richard G. 2016. *How Men Age: What Evolution Reveals about Male Health and Mortality*. Princeton, NJ: Princeton University Press.

Buttolph, Lita P., William Kay, Susan Charnley, Cassandra Moseley, and Ellen M. Donoghue. 2006. "Northwest Forest Plan—The First 10 Years (1994–2003): Socioeconomic Monitoring of the Olympic National Forest and Three Local Communities." *Pacific Northwest Research Station General Technical Report*, PNW-GTR-679:84. Portland, Oregon.

Charnley, Susan, Delilah Jaworski, Heidi Huber-Stearns, Eric M. White, Elisabeth Grinspoon, Rebecca J. McLain, and Lee Cerveny. 2018. "Chapter 10—Environmental Justice, Low-Income and Minority Populations, and Forest Management in the Northwest Forest Plan Area." In *Synthesis of Science to Inform Land Management Within the Northwest Forest Plan Area (General Technical Report PNW-GTR-966)*, edited by T. A. Spies, P. A. Stine, R. Gravenmier, J. W. Long, and M. J. Reilly, 809–849. Portland, OR: US Forest Service.

Charnley, Susan, Rebecca J. McLain, and Ellen M. Donoghue. 2008. "Forest Management Policy, Amenity Migration, and Community Well-Being in the American West: Reflections from the Northwest Forest Plan." *Human Ecology* 36:743–761. https://doi.org/10.1007/s10745-008-9192-3.

Colfer, A. Michael, and Carol J. Pierce Colfer. 1979. *Big Money, Small Change: Cultural Paradigms in an American Village*. Cambridge, MA: Abt Associates.

Colfer, Carol J. Pierce. 1977. *Women's Communication and Family Planning in Rural America: The Case of Bushler Bay*. Edited by D. Lawrence Kincaid. Vol. 4, *Case Studies in Family Planning*. Honolulu, HI: EWCI (Case Study #4; reprinted 1978).

Colfer, Carol J. Pierce. 1985a. "Female Status and Action in Two Dayak Communities." In *Women in Asia and the Pacific: Toward an East-West Dialogue*, edited by M. J. Goodman, 183–214. Honolulu: University of Hawaii Press.

Colfer, Carol J. Pierce. 1985b. "On Circular Migration: From the Distaff Side." In *Labour Circulation and the Labour Process*, edited by G. Standing, 219–251. London: Croom Helm.

Colfer, Carol J. Pierce. 2018. "Before and After the 'Timber Wars': Context, Change and Potential Collaboration on the Olympic Peninsula." *Humboldt Journal of Social Relations* 40:137–162.

Colfer, Carol J. Pierce, Lee Cerveny, and Susan Stevens Hummel. 2019. "Using Rapid Rural Appraisal Tools to Explore Gender and Forests in the Global North." *Human Organization* 78 (1):12–27.

Colfer, Carol J. Pierce, Iliana Monterroso, and Markus Ihalainen. (2020, in press). "Understanding Gender Dynamics in the Context of Rural Transformation Processes: A Case from East Kalimantan." *CIFOR Occasional Paper*.

Colfer, Carol J. Pierce, and Juk Along Pelibut. 2001. "Beadlore of the Uma' Jalan Kenyah." *The Sarawak Museum Journal* 77 (new series) (LVI):29–35.

Colfer, Carol J. Pierce, and Herwasono Soedjito. 2003. "Food, Forest, and Fields in a Bornean Rain Forest: Toward Appropriate Agroforestry Development." In *Borneo in Transition: People, Forests, Conservation and Development*, edited by Christine Padoch and Nancy Peluso, 215–223. Kuala Lumpur: Oxford University Press. Original edition, 1996.

Connell, R. W. 1995. *Masculinity*. Berkeley, CA: University of California Press.

Cornwall, Andrea. 2016. "Introduction: Masculinities under Neoliberalism." In *Masculinities under Neoliberalism*, edited by Andrea Cornwall, Frank G. Karioris, and Nancy Lindisfarne, 1–28. London: Zed Books.

Dark, A. 1997. "Landscape and Politics on the Olympic Peninsula: Social Agendas and Contested Practices in Scientific Forestry." *Journal of Political Ecology* 4 (1):1–25.

Davis, Emily Jane, Reem Hajjar, Susan Charnley, Cassandra Moseley, Kendra Wendel, and Meredith Jacobson. 2020. "Community-based Forestry on Federal Lands in the Western United States: A Synthesis and Call for Renewed Research." *Forest Policy and Economics* 111:1–12.

Elias, Marlène, Netsayi Mudege, Diana E. Lopez, Dina Najjar, Vongai Kandiwa, Joyce Luis, Jummai Yila, Amare Tegbaru, Gaya Ibrahim, Lone Badstue, Esther Njuguna-Mungai, and Abderahim Bentaibi. 2018. "Gendered Aspirations and Occupations Among Rural Youth, in Agriculture and Beyond: A Cross-regional Perspective." *Journal of Gender, Agriculture and Food Security* 3 (1):82–107.

Elmhirst, Rebecca, Mia Siscawati, and Carol J. Pierce Colfer. 2016. "Revisiting Gender and Forestry in Long Segar, East Kalimantan: Oil Palm and Divided Aspirations." In *Gender and Forestry: Climate Change, Tenure, Value Chains, and Emerging Issues*, edited by Carol J. Pierce Colfer, Bimbika Sijapati Basnett, and Marléne Elias, 300–317. London: Earthscan/CIFOR.

Fisher, Micah R., Timothy Workman, Agus Mulyana, Balang Institute, Moira Moeliono, E. Linda Yuliani, Carol J. Pierce Colfer, and Urban El Fatih Bani Adam. 2017. "Striving for PAR Excellence in Land Use Planning: Multi-Stakeholder Collaboration on Customary Forest Recognition in Bulukumba, South Sulawesi." *Land Use Policy*. http://dx.doi.org/10.1016/j.landusepol.2017.09.057.

Gaveau, David L. A., Bruno Locatelli, Mohammad A. Salim, Husna Yaen, Pablo Pacheco, and Douglas Sheil. 2018. "Rise and Fall of Forest Loss and Industrial Plantations in Borneo (2000–2017)." *Conservation Letters* 8:1–8. wileyonlinelibrary.com/journal/conl.

Hatch, Laurie Russell. 2000. *Beyond Gender Differences: Adaptation to Aging in Life Course Perspective*. London: Routledge.

Hayns, Joe. 2016. "Desperate Markets and Desperate Masculinities in Morocco." In *Masculinities under Neoliberalism*, edited by Andrea Cornwall, Frank G. Karioris, and Nancy Lindisfarne, 99–110. London: Zed Books.

Jackson, Cecile. 2001. "Men at Work." In *Men at Work: Labour, Masculinities, Development*, edited by Cecile Jackson, 1–22. Portland, OR: Frank Cass, with EADI (European Association of Development Research and Training Institutes, Bonn).

Kirschner, Annabel R. 2010. "Understanding Poverty and Unemployment on the Olympic Peninsula after the Spotted Owl." *The Social Science Journal* 47 (2):344–358.

Manurung, Butet. 2019. "'Normalising' the Orang Rimba." In *Contentious Belonging: The Place of Minorities in Indonesia*, edited by Greg Fealy and Ronit Ricci, 232–251. Singapore: ISEAS Yusof Ishak Institute.

Marfo, Emmanuel, Carol J. Pierce Colfer, Bocar Kante, and Silvel Elias. 2010. "From Discourse to Policy: The Practical Interface of Statutory and Customary Land and Forest Rights." In *Forest for People: Community Rights and Forest Tenure Reform*, edited by Anne Larson, Deborah Barry, Ganga Ram Dahal, and Carol J. Pierce Colfer, 69–89. London: Earthscan/CIFOR.

McLain, R. J., and Lynch, K. 2010. "Managing Floral Greens in a Globalized Economy: Resource Tenure, Labour Relations, and Immigration Policy in the Pacific Northwest, USA." In *Wild Product Governance: Finding Policies that Work for Nontimber Forest Products*, edited by S. A. Laird, R. J. McLain, and R. P. Wynberg, 265–286. Washington, DC: Earthscan.

Nader, Laura. 1972. *Up the Anthropologist: Perspectives Gained From Studying Up*. Washington, DC: US Department of Health, Education and Welfare.

Peluso, Nancy Lee. 2008. "A Political Ecology of Violence and Territory in West Kalimantan." *Asia Pacific Viewpoint* 49 (1):48–67.

Petesch, Patti, Renee Bullock, Shelley Feldman, Lone Badstue, Anne Rietveld, Wenda Bauchspies, Adelbertus Kamanzi, Amare Tegbaru, and Jummai Yila. 2018. "Local Normative Climate Shaping Agency and Agricultural Livelihoods in Sub-Saharan Africa." *Journal of Gender, Agriculture and Food Security* 3 (1):108–130.

Pollock, David C., and Ruth Van Reken. 2001. *Third Culture Kids: The Experience of Growing Up Among Worlds*. Boston: Nicholas Brealey Publishing.

Reed, Maureen G. 2003. "Marginality and Gender at Work in Forestry Communities of British Columbia, Canada." *Journal of Rural Studies* 19:373–389.

Roberts, Kathleen Glenister. 2011. "War, Masculinity, and Native Americans." In *Global Masculinities and Manhood*, edited by Ronald L. Jackson II and Murali Balaji, 141–160. Urbana, IL: University of Illinois Press.

Santika, Truly, Kerrie A. Wilson, Sugeng Budiharta, Elizabeth A. Law, Tun Min Poh, Marc Ancrenaz, Matthew J. Struebig, and Erik Meijaard. 2019. "Does Oil Palm Agriculture Help Alleviate Poverty? A Multidimensional Counterfactual Assessment of Oil Palm Development in Indonesia." *World Development* 120:105–117.

"Securing Customary Rights to Forests in Maluku Province, Indonesia". 2020. Bogor, Indonesia: CIFOR. www.cifor.org/gcs-tenure/.

Siscawati, Mia. 2020. "Gender and Forest Tenure Reform in Indonesia." *CIFOR Working Paper* 258, Bogor, Indonesia.

Spiller, Henry. 2010. *Erotic Triangles: Sundanese Dance and Masculinity in West Java*. Chicago, IL: University of Chicago Press.

Tsing, A. L. 2015. *The Mushroom at the End of the World: On the Possibility of Life in Capitalist Ruins*. Princeton, NJ: Princeton University Press.

Varley, Ann, and Maribel Blasco. 2001. "Exiled to the Home: Masculinity and Ageing in Urban Mexico." In *Men at Work: Labour, Masculinities, Development*, edited by Cecile Jackson, 115–138. Portland, OR: Frank Cass with EADI (European Association of Development Research and Training Institutes, Bonn).

Vera-Sanso, Penny. 2016. "Taking the Long View: Attaining and Sustaining Masculinity Across the Life Course in South India." In *Masculinities Under Neoliberalism*, edited by Andrea Cornwall, Frank G. Karioris, and Nancy Lindisfarne, 80–98. London: Zed Books.

8

THE RELEVANCE OF MASCULINITIES FOR FORESTS AND THEIR MANAGEMENT

Introduction

I began this book with my own personal experience of masculinity throughout the US and globally, in earlier times. The 'meat' of the book has been the four cases: *Bushler Bay* in the rural US (mid-1970s), *Long Segar* in rural Kalimantan (late 1970s and beyond), *Sitiung* in rural Sumatra (mid-1980s) and *elite forestry researchers* globally (1990s and 2000s), with three more recent updates in Chapter 7. From these cases, analyzed in depth, I have demonstrated some of the variety that exists globally in forest masculinities and some of the varying choices and preferences that men in these forests have made regarding their own identities. Here I turn to a summary of the forest implications of these variations in masculinities.

Forest relevance

> Feminists were interested not just in explaining how sciences worked, leaving their projects and practices unchanged. . . . Instead, *feminists aimed to change scientific practice, to produce empirically and theoretically more successful research*. . . . Moreover, as feminist critics of the First World's development policies in the Third World began to examine the destructive consequences of the imposition in the Third world of First World scientific and technological assumptions and practices, it became clear that far deeper and broader changes in scientific practice and philosophies of science would be required if sciences were to speak also for the 70 percent or so of the world's most economically and politically vulnerable women, men and children . . .
>
> *(Harding 2004, p. 31; my emphasis)*

This has indeed been part of my motivation in writing this book. I have highlighted the differences among communities—specifically in the ways they viewed and practised their masculinities—in and near different forests. And I have hoped, by bringing these differences to light, to influence the ways forestry science and 'development' are practised. But to do that, there remains another task: to show how these differences relate specifically to forest management policies and practices. That is my goal here.

Understanding men, as individuals central to global views of forests, is fundamental if we seek styles of forest management that benefit both environments and their inhabitants. Environments and cultures (integrally related to being human) vary enormously. Accordingly, men's relations with forests vary. Good management takes into account the knowledge, capabilities, needs and goals of the people who live in and near such forests. Taking these differences into account becomes even more important as we strive to work collaboratively with local people to maintain forests and forest cultures and to improve forest management generally. We need to become attuned to the impacts of power differences—particularly the power we ourselves often inadvertently wield—in our interactions with communities (see Rosendahl et al. 2015, who argue for what they call 'strong objectivity', for a nice analysis of one common and relevant and inadvertently marginalizing scenario).

Like these authors, we need to consider such issues from a transdisciplinary perspective. In this book, I have used my own methodological and analytical tools, informed both by my anthropological training and by decades of work with foresters, but I recognize that were I to collaborate with a forester, he or she might come up with additional ways that these social facts bear on forests and vice versa. A forester might be able to suggest specific and practical ways to alter management that would have allowed us to avoid or moderate the dislocations discussed in some of the cases below—an endeavour that remains for future analyses.

In the following sections, I consider the forest implications of the specific masculinities described in this book. The trajectories over time may be helpful to us in anticipating changes in the future and developing more effective and benign forest management strategies.

Forest relevance of Bushler Bay masculinities

As the timber wars began (1970s)

The implications of forest management were quite different for Local men vis-à-vis Public Employee men in Bushler Bay, though both were affected by the community's strong gender polarity. Men did not want to be like women and most went to considerable effort to ensure that they were not. For Local men, this meant emphasizing their involvement in an occupation that occurred outdoors and required physical strength, endurance and courage. Logging represented a key chord in achieving manliness and a crucial tone in most loggers' songs. For Public Employee men, symbolized within the community by US

Forest Service employment, secure breadwinning was the central tone they plucked.

US government policies about forest management—sparked by concerns about the spotted owl and the marbled murrelet—had different implications for the men who plucked these divergent harp strings. Reductions in timber sales by the US Forest Service (USFS) in this area where most forests were indeed federally owned meant loss of income, but more importantly for Local men, loss of opportunities to fulfill their most cherished masculine ideals (physical strength, outdoor work, love of nature, independence). Such loss was a direct attack on their feelings about their manhood and sense of self.

For Public Employee men, there were similar though less extreme, dislocations. The reduction in timber sales meant a reduction in staffing of the local USFS office and resulting uncertainty about continued employment as well (the breadwinning harp string so central to Public Employee men).

Compounding these disharmonies was the wider American emphasis on competition—part of the harp itself. Competition among both Locals and Public Employees increased as forest-related economic opportunities diminished for each, with a corresponding diminution of community cooperation. Anger and distrust among Locals, much of which was directed toward the Public Employees who had to enforce the national laws, grew, exacerbated by feelings of powerlessness as their voices carried no weight in the federal decision-making process that so harshly affected them.[1] Locals felt they were fighting for their way of life, and they were losing. Public Employees, though also unhappy, were more likely to have employment options in other parts of the country, moving, but remaining employed, secure, with employment benefits—and thus their masculinity intact. Anger, dislocation and anomie pervaded the community.

The harp—the cultural and institutional framework—represented in Bushler Bay carried with it other losses as well. The knowledge that these men of the forest held, from long and intimate experience in it, was lost to the US Forest Service and its formal managers. There were no meaningful feedback channels for local information, knowledge, wisdom to go 'up' the bureaucratic ladder to those who made the decisions about forests.

Additionally, the traditional emphasis on timber, and the assumptions by formal forest managers that forests were masculine places, resulted in even less attention to women's forest knowledge, practices and hopes for the future. At that time, this may have been an advantage for rural women, as the USFS paid little attention to subsistence uses (berries, mushrooms, ferns and other NTFPs), and both men and women could manage and harvest, consistent with community norms. But it was a loss again for the USFS of local knowledge about these resources.

Finally, the lack of knowledge by USFS employees about Local lifeways and the likely impacts of federal forest management actions eventually represented a loss of cultural diversity as urbanites 'invaded'—and a loss to humanity.

- Long Segar's muted gender differentiation vs. the polarized gender of Bushler Bay
- Kenyah traditions emphasizing headhunting, swidden agriculture and routine dependence on a vast repertoire of forest products (see Colfer 2008) vis-à-vis Bushler Bay's American frontier orientation and dependence most fundamentally on logging and efficient tree production and harvest (Colfer and Colfer 1978; Colfer 2018)

Given the very mild gender differentiation, there was little felt need among Kenyah men to focus much attention on demonstrating their masculinity. Thus, changes to their forest-dependence were not as much of a direct attack on their manhood as in Bushler Bay. These changes did however represent an attack on their ethnic identities, which were firmly based in the practice of swidden agriculture. Swidden, in turn, was closely linked to women's gender identities (also less key to women's sense of self than among Bushler Bay's men and, to a lesser extent, women).

The forested environment—a mosaic of primary and secondary forest, new and older swiddens, as well as the village itself—provided men the opportunity to gain and demonstrate forest-related skills, like hunting, fishing, gathering forest products and finding their way through dense forest. Demonstration of such skills represented one path for men to persuade desirable and hardworking women to marry them, gain and maintain the respect of their families and potentially become a village leader in time.

The community was strongly and directly dependent on the forest mosaic as a source of many products, which men (and women) obtained for daily life (timber, firewood, poles, NTFPs, medicinal plants, fibres and almost all food). Even the rice produced in swiddens was dependent on the forest; the soils were poor and required the cycle of cutting, agricultural use and fallow periods of forest regrowth to maintain fertility, a form of 'restoration' not usually recognized. The dangers in the forest—wild animals, accidents, losing one's way, even antagonistic people—served as context for demonstrating courage, strength and knowledge acquisition. These harp strings were also available to and admired for those (fewer) women who plucked such strings (e.g., Box 7.5).

Formal forest managers, the most evident being Americans working at the Georgia Pacific Timber concession, saw local people as potential problems, related to land tenure, labour and potential conflict. Their focus, as in Bushler Bay, was on production, 'getting the logs out'. Like the USFS in Bushler Bay, they had little interest in the community beyond preventing problems and their escalation.

However, there were ways in which closer links between the communities and the companies or the Indonesian Forest Service could have been of mutual benefit. Local people had deep understanding of local forests. Their knowledge would have been more compatible with a style of management now more in vogue, one that recognized the multiple values extant in a forest, beyond just commercially valuable trees.

An attitude of mutual respect and openness to the potential for valuable inputs from these Kenyah men (and women) would have helped logging companies and government managers manage the forest in ways that were more commercially, ecologically and socially sustainable. A greater variety of commercially valuable products could have been identified with local help, and together mechanisms for managing them sustainably could have been developed.

- Local knowledge of wildlife patterns of behaviour and reproduction could have led to management of wildlife for continued use and contribution to local diets while protecting those as needed (e.g., orangutans and other endangered species).
- Non-timber forest products, like rattan, bamboo, *sang* (*Licuala spp.*) and a huge variety of fibres and edible plants, which had subsistence and potential commercial value, could have been studied. Management plans could have been developed collaboratively such that local people and/or companies could harvest them sustainably.
- Medicinal plants were recognized by the Kenyah (Leaman 1996), some of which could well have been commercially valuable—building on local knowledge, as pharmaceutical companies have done elsewhere.

These kinds of interactions between forestry professionals, whether governmental or in private industry, could have prevented the sad situation we now face: The forest is gone, basically. This amazing forest, of value to all of humanity, has now been replaced by a 'forest' of oil palm, with only scrubby remnants of the original forest along the river banks, still used for swiddens, but with unsustainably reduced fallow periods due to lack of access to land (now in the hands of oil palm companies). This change is due far more to government and industry action than to the swiddeners so often blamed (and who co-existed with the forest for centuries).

But there is room for optimism if formal forest managers could be persuaded to work collaboratively with those who remain in the dwindling forested areas. Much of the people's ecological knowledge remains, for the moment, and saving forest-based cultures should be as much of a priority as saving biodiversity.

Kenyah swiddeners in fields of oil palm

My own sorrows at the loss of the Long Segar forest are clear, and to some extent shared by the Kenyah. However, their perspectives are more optimistic than mine. They are a remarkably adaptable people.

Amazingly, gender differentiation remains muted, despite increasing pressure to conform to longstanding global narratives of masculine superiority. As 'modernity' has made its way into Long Segar,[5] the churches, mosques, government and to a lesser degree, industry, all sing the song of men's superiority over women.

The Kenyah's near-total dependence on the forest has disappeared with the forest itself. Unlike previously, many foods are bought now. Most forest products are no longer available in amongst the oil palms. Kenyah men can no longer demonstrate their strength, fortitude and courage by manoeuvring through the nearby forest, thus obliterating previously important masculine harp strings, avenues to masculine identity.[6] Trade goods, the most obvious motivation for men to make the expeditions traditionally associated with manhood, are now available locally, removing or at least altering the expedition-making harp string previously so valued.

Denied their role during recent decades as independent or contract loggers by the absence of forests others wanted cleared, Kenyah men are taking up small-scale, local plantation development in concert with the powerful industry actors nearby—strongly preferring independent or semi-independent agroforestry production to wage labour. Where women have taken part-time jobs with the oil palm companies, fewer men have been so inclined.

With only small areas of forest remaining along the river's edge, men's links to forests are disappearing in a process over which they have little control. Yet their sense of adventure and curiosity (and perhaps the flexibility of their own ideas about masculinity) appear to grant them an equanimity in the face of drastic social, cultural and environmental change that was less evident in the Bushler Bay example.

The oil palm industry has shown a willingness to work with communities, though often in a manipulative manner—building on existing intra-community antipathies and exacerbating distrust and factionalism in their attempts to obtain additional land for their plantations. Their presence, much closer geographically to the community than the timber company offices had been, provides examples, knowledge of oil palm cultivation, that local people can follow, and the companies provide infrastructural improvements sometimes (e.g., paving village roads).

I saw no evidence in 2019 of formal Indonesian forestry involvement. But there would be opportunities to work collaboratively to improve the management of the strips along the riverside, perhaps restoring elements of the original forests. Forestry professionals could work with the Kenyah to resuscitate some of the forest products used previously for sale, for subsistence and/or for artistic expression and even for ecological purposes. See e.g., Gibson and Warren (in press), on the efforts of individual men in Australia, Hawaii and the Pacific Northwest of the US, trying to establish the trees that can in the distant future provide the raw materials, now rare, to make fine guitars. Forestry professionals could help link people with, or develop, trade networks[7] to reinforce such plantings and encourage wildlings, including tree species like ironwood (*Eusideroxylon zwageri*) or 'honey trees' (*Koompassia excelsa*)—though Indonesian laws granting rights to all wild trees to the Indonesian government would have to change.[8] The people and the forestry professionals could together figure out agroforestry strategies that would benefit both the forest and the people, including strip plantings, species needing shade, modernized agroforestry and other well-known techniques.

These actions would not replace the forest or its functions either ecologically or socially, including in terms of masculinities. But they could contribute to biodiversity conservation and improve local people's livelihoods and important aspects of their forest-based cultures. Such interaction could strengthen beneficial adaptation of externally planned and implemented policies with profound implications for people's lives and local contexts. The simple act of respecting the knowledge of rural men (and women) would also go far in making Indonesian forest (and other kinds of) management more coherent.

Forest relevance of Sitiung masculinities

Sadly, I have not been able to return to Sitiung since 2008, to see how masculinities may have changed. However, my last brief visit revealed a landscape covered primarily in oil palm rather than forest. Google Earth shows that the process of conversion has continued apace, as have a series of interviews conducted by Ardi et al. (2020).

Gender differentiation among the three ethnic groups in Sitiung—Minang, Javanese and Sundanese—took intermediate positions between the two extremes of Bushler Bay and Long Segar. The histories of all three resident ethnic groups included dynasties, sultanates or kingdoms ('Great Traditions', à la Redfield 1960) and all three adhered, to varying degrees, to Islamic narratives of masculine superiority.[9] All of these communities resided in forested areas in the 1980s, but none would have been considered truly 'forest-based cultures', though all made use of nearby forests.

Masculinities for the Minang focused on cleverness, articulateness, religiosity and making money. Their moneymaking was as likely to involve forest use (often clearing for rice and then rubber or other tree crops) as any other profitable activity. But their masculinities were far less dependent on the forest than those of the Kenyah or Bushler Bay's loggers.

Javanese men (and women) have tended to fear the forest, thus reducing further any links between masculinities and forests. Instead, men's focus tended to be on agriculture and fulfilling the expectations of their place in society (which they as farmers generally perceived to be quite low). In Sitiung 5, men logged, because it was necessary to clear their fields and because they needed money to survive, but it was not their choice; it was comparatively minimally related to demonstration of their masculinity (only as evidence of hard work and an under-valued version of breadwinning). In the longer established Sitiung 1, men had even fewer links with the forest. Indeed, each family was supposed to have a second hectare designated for rice, but families were reluctant to go out into the forest, figure out which land belonged to them and clear it, satisfied instead to focus on their bulldozed and infertile land nearer the community (labour constraints were also a prime factor).

Many Sundanese shared with the Minang concern with religiosity and making money and with the Javanese, a concern with agricultural production. Many, recognizing the minimal opportunities in Sitiung 5 to make money or to succeed at fishponds, rice or vegetable production, returned to West Java. Rice they deemed more central to their well-being than did the Javanese, who were willing to eat corn when rice crops failed.

The implications of these views of masculinity reduce the options for formal forest managers' involvement with the communities. The amount of available, traditional knowledge of the forest is less than in Long Segar or Bushler Bay, where forests form important parts of people's identities.

The Indonesian Forest Service's interests (long dominated by the Javanese) have tended to focus on replanting/regreening, which could be emphasized among the Javanese and Sundanese, as there would be little need for these farmers to spend time in the (now nearly non-existent) forest per se. The willingness of the Javanese to comply with bureaucratic dictates also opens doors for top-down planning that is less likely to succeed for the culturally more independent Minang, for instance. Minang interests in trees have been in those with commercial value, so that interest could also be built upon, including forest trees, as the Minang showed little fear of the forest per se (excepting tigers, still a danger in the 1980s). Again, as noted for Long Segar, the legal ownership of wild trees by the government would have to shift to grant community members rights to own such trees. These ethnic groups were more admired within Indonesia and would therefore encounter fewer prejudices and fears from forestry officials, compared to the Kenyah or other Dayaks. This in turn would make collaboration easier. The fact that Ardi and associates (2020) found Sitiung's people to have certificates of ownership of their land (a big change from the past) suggests this may be correct.

The implications of forest loss are also less extreme for these ethnic groups, as their cultural systems can function adequately and adapt with or without forests.

Forest relevance of forestry scientists' masculinities

There is a range among forestry scientists in the strength of their gender differentiation, influenced by their native cultures as well as that of their work setting. International work settings are likely to include at least lip service and sometimes genuine commitment to global efforts to enhance gender equity. Gender equity does not theoretically *require* either muted or strong differentiation between men and women; both can probably be equitable.

Although scientists typically recognize and value the masculinities described for a place like Bushler Bay (e.g., physical strength, ability to endure harsh conditions), these are less likely to be the masculinities to which they themselves aspire. Scientists (men) are more likely to pluck the harp strings of adventure and articulateness in speech (as do the Kenyah men of Long Segar), logic in thought (like Minang men) and quantification in analysis (like Americans; what Clarke and Hamilton 2013 call the 'magic counting dragon').[10]

The global narrative of forests as masculine places provides a convenient rationale for these scientists' traditional focus on timber, the product virtually universally associated with men. However, global efforts both to recognize the forest as a complex biome with values beyond timber and to improve gender equity have strengthened these men's attention to forest values other than timber. These in turn have reinforced research on the norms, roles, activities and goals of women as well as men—sometimes leading to a muting of gender differentiation and a broader understanding of forest uses.

The dominance (both numerically and substantively) of men in formal forestry has more commonly resulted in standardized policies designed and implemented with their own interests, concerns and knowledge in mind. They often have access to little understanding of rural forest masculinities or their variability, which has interfered with their ability to tailor their advice, research and planning to rural realities. The variations in rural masculinities have suffered a similar 'invisibility' to that of women in forests (e.g., Arora-Jonsson, González-Hidalgo, and Colfer 2019; Paulson 2017).

Some of the harp strings these scientists pluck for themselves can function to marginalize others. The emphasis of some on control suggests an unwillingness to accept control that local people might legitimately wish to exercise; the interest in demonstrating articulateness can mean a lack of willingness to listen, as is necessary if we want to understand local perspectives.

Making the links called for here will require the involvement of such scientists, researchers and professional foresters. They will benefit from

- Examining and recognizing their own assumptions and being open to seeing alternative masculinities
- Willingness to consider qualitative understandings as legitimate complements to the more familiar quantitative approaches
- Bringing pressure to bear on forest institutions like the US Forest Service or Indonesia's forest bureaucracy to increase their flexibility and their responsiveness to local variation
- Developing and expressing respect for and openness to knowledge that local people have about forests, combined with a willingness to share their own knowledge and approaches in two-way communication
- Monitoring these efforts and adapting those that fail with perseverance

Last words

I hope that I have succeeded in demonstrating in this book, via this shared journey, some of the variations that characterize masculinities and their differing links with forests in different places and at different times. Such links can range from full and thriving to absent, and many levels of involvement between those two extremes.

The implications of such variation for forest management are significant. We are concerned to enhance the viability of forest management for the forests themselves (and the broader environment) and also for the well-being of the human populations who reside in and near those forests. Forests have been widely defined as masculine settings, based largely on the hegemony of the definitions and narratives coming from the global North, many of which differ markedly from those of forest peoples. This needs to change if we have any hope to keep the world's forests for coming generations.

Notes

1 One can easily see the parallels with communities in the global South.

2 In my 2017 re-visit, for instance, I contacted the director ahead of time about the planned research on gender, community and forests, worked closely with USFS research personnel in their area (Portland, OR and Seattle, WA), and had the prestige of Cornell University and the Center for International Forestry Research behind me. Yet to my amazement, no one showed up for the meeting they had scheduled for me. After several attempts, the director agreed reluctantly to spend 10–15 minutes with me, before rushing off to take care of her home repairs. I was able to talk with a couple of employees informally, but one called afterwards to make doubly certain the opinions expressed would remain confidential, fearing adverse responses from supervisors. These are not the actions of an institution interested in local communities or gender. A certain paranoia is likely based on the longstanding conflict.

3 I heard statements like, "I don't mix with the community", from several newcomers and longer-resident Public Employees.

4 One encouraging possibility exists in the USDA's Collaborative Forest Landscape Restoration Program; see e.g. Butler and Schultz 2019; Walpole et al. 2017; Wagner and Fernandez-Gimenez 2008.

5 Even some of the oldest people have and use cell phones, as they sit in traditional fashion on their front porches, watching the world go by. Young people use Facebook regularly, keeping in touch with their relatives all over East Kalimantan (a few in Jakarta and at least one anthropologist in the US). There are computers, flash drives and Wi-Fi available in the village office, and there's now a small library.

6 Several Kenyah men and I got lost in the exactly replicated matrix formed by the local oil palm plantations in March 2019. They were as clueless as I about how to find our way out.

7 Such collaboration was successfully undertaken in a number of countries within the Adaptive Collaborative Management Program at CIFOR (see www.cifor.org/acm/).

8 See https://forestlegality.org/risk-tool/country/indonesia (accessed 16 August 2019), for a sense of the complexity of Indonesian forest management laws and regulations. Larson et al. (2019) document the reform efforts (and level of success) focused on community use of forests.

9 Remembering of course that Indonesian societies are nowhere near the extreme gender differentiation of Middle Eastern versions of Islam or masculinities.

10 "The 'MCD' [magic counting dragon] is a symbol representing the cultural forces that confer a kind of security and legitimacy in human activity. In an era of scientific inquiry marked by what has been called 'the statistical style' . . . numbers are a central feature of our magic system . . ." (p. 31).

References

Ardi, with Zumroatun Nikmah, Nori Helpita, Arjunawati, Siska Purnama Sari, Zulkifli, Alprian Apred S., and Rahmat Hidayat. 2020. *Hasil Survei Sitiung 2020*. Report Generated for Colfer in Response to Her Questions about Sitiung Currently.

Arora-Jonsson, Seema, Marien González-Hidalgo, and Carol J. Pierce Colfer. 2019. "Towards a Future Forest (and Feminist) Agenda: Acknowledging 'the Invisible' in Forestry Relations." *Human Ecology*, submitted to.

Brandth, Berit, and Marit S. Haugen. 2000. "From Lumberjack to Business Manager: Masculinity in the Norwegian Forestry Press." *Journal of Rural Studies* 16:343–355.

Butler, William H., and Courtney A. Schultz, eds. 2019. *A New Era for Collaborative Forest Management: Policy and Practice Insights from the Collaborative Forest Landscape Restoration Program*. London: Earthscan/Routledge.

Clarke, Paula K., and W. Ted Hamilton. 2013. "The Value of Anthropology: Noticing and Challenging the Magic Counting Dragon." *Anthropology News*:31.

Colfer, Carol J. Pierce. 2008. *The Longhouse of the Tarsier: Changing Landscapes, Gender and Well Being in Borneo*. Edited by G. N. Appell and Vinson H. Sutlive. Vol. 10, *Borneo Research Council Monograph Series*. Phillips, ME: Borneo Research Council, in cooperation with CIFOR and UNESCO.

Colfer, Carol J. Pierce. 2018. "Before and After the 'Timber Wars': Context, Change and Potential Collaboration on the Olympic Peninsula." *Humboldt Journal of Social Relations* 40:137–162.

Colfer, Carol J. Pierce, Lee Cerveny, and Susan Stevens Hummel. 2019. "Using Rapid Rural Appraisal Tools to Explore Gender and Forests in the Global North." *Human Organization* 78 (1):12–27.

Colfer, Carol J. Pierce, with A. Michael Colfer. 1978. "Inside Bushler Bay: Lifeways in Counterpoint." *Rural Sociology* 42 (2):204–220.

Gibson, Chris, and Andrew Warren. in press. "Keeping Time with Trees: Climate Change, Forest Resources, and Experimental Relations with the Future." *Geoforum*, Vol. 108, January 2020, 325–337. https://doi.org/10.1016/j.geoforum.2019.02.017.

Harding, Sandra. 2004. "A Socially Relevant Philosophy of Science? Resources from Standpoint Theory's Controversiality." *Hypatia* 19 (1):25–46.

Larson, Anne M., Iliana Monterroso, Nining Liswanti, Tuti Herawati, Abwoli Banana, Pamela Cantuarias, Karin Rivera, and Esther Mwangi. 2019. "Models for Formalizing Customary and Community Forest Lands: The Need to Integrate Livelihoods into Rights and Forest Conservation Goals." *CIFOR InfoBrief* 253:8. https://doi.org/10.17528/cifor/007273.

Leaman, D. J. 1996. "The Medicinal Ethnobotany of the Kenyah of East Kalimantan (Indonesian Borneo)." Dissertation, University of Ottawa.

Paulson, Susan. 2017. "Changing Masculinities and Femininities in the (Re)Production of Andean Silvopastoral Systems." In *The Earthscan Reader on Gender and Forests*, edited by Carol J. Pierce Colfer, Marlène Elias, Bimbika Sijapati Basnett, and Susan Stevens Hummel, 163–280. London: Earthscan/Routledge.

Redfield, Robert. 1960. *Peasant Society and Culture*. Chicago, IL: University of Chicago Press.

Rosendahl, J., M. A. Zanella, S. Rist, and J. Weigelt. 2015. "Scientists' Situated Knowledge: Strong Objectivity in Transdisciplinarity." *Futures* 65:17–27. http://doi.org/10.1016/j.futures.2014.10.011.

Singh, Neera M. 2013. "The Affective Labor of Growing Forests and the Becoming of Environmental Subjects: Rethinking Environmentality in Odisha, India." *Geoforum* 47:189–198.

Wagner, Cheryl L., and Maria E. Fernandez-Gimenez. 2008. "Does Community-Based Collaborative Resource Management Increase Social Capital?" *Society and Natural Resources* 21:324–344. https://doi.org/10.1080/08941920701864344.

Walpole, Emily H., Eric Toman, Robyn S. Wilson, and Melanie Stidham. 2017. "Shared Visions, Future Challenges: A Case Study of Three Collaborative Forest Landscape Restoration Program Locations." *Ecology and Society* 22 (2). https://doi.org/10.5751/ES-09248-220235.

INDEX

Note: Page numbers in *italics* indicate figures and those in **bold** indicate tables.

For Product Safety Concerns and Information please contact our EU representative GPSR@taylorandfrancis.com
Taylor & Francis Verlag GmbH, Kaufingerstraße 24, 80331 München, Germany

www.ingramcontent.com/pod-product-compliance
Ingram Content Group UK Ltd.
Pitfield, Milton Keynes, MK11 3LW, UK
UKHW022224190926
475702UK00003B/487
* 9 7 8 0 3 6 7 4 0 3 8 9 8 *